Wine Sales and Distribution

The Secrets to Building a Consultative Selling Approach

Paul Wagner, John C. Crotts,
and Byron Marlowe

ROWMAN & LITTLEFIELD
Lanham • Boulder • New York • London

Published by Rowman & Littlefield
An imprint of The Rowman & Littlefield Publishing Group, Inc.
4501 Forbes Boulevard, Suite 200, Lanham, Maryland 20706
www.rowman.com

6 Tinworth Street, London SE11 5AL, United Kingdom

British Library Cataloguing in Publication Information Available

Library of Congress Cataloging-in-Publication Data
Names: Wagner, Paul, 1952- author. | Crotts, John C., author. | Marlowe, Byron, author.
Title: Wine sales and distribution : the secrets to building a consultative selling approach / Paul Wagner, John C. Crotts, and Byron Marlowe.
Description: Lanham : Rowman & Littlefield, [2019] | Includes bibliographical references and index.
Identifiers: LCCN 2019005090 (print) | LCCN 2019007368 (ebook) | ISBN 9781538117323 (ebook) | ISBN 9781538117309 (cloth : alk. paper) | ISBN 9781538117316 (pbk. : alk. paper)
Subjects: LCSH: Wine industry—United States. | Wine—Marketing. | Sales management.
Classification: LCC HD9375 (ebook) | LCC HD9375 .W343 2019 (print) | DDC 663/.200688—dc23
LC record available at https://lccn.loc.gov/2019005090

Contents

List of Figures and Tables   v

Preface   vii

Acknowledgments   xi

SECTION I: PROFESSIONAL FOUNDATIONS

1   Wine Sales: Foundations of Success   3

2   Buyer-Supplier Relationships and Compliance Regulations in the Wine Industry   13

3   The Exchange of Value between Buyers and Sellers   23

4   The Organization of a Sales Force   39

5   Direct-to-Consumer Sales   51

SECTION II: THE CONSULTATIVE SALES PROCESS

6   Buyer Motivations and Presales Call Planning   67

7   Precall Research   77

8   Call Opening   87

9   Probing the Customer's Needs   97

10   Supporting the Needs of Your Customers   109

11   Closing the Sale   125

12   Negotiating Customer Concerns   137

13   Selling to a Lack of Interest   153

SECTION III: WINE TRADE

14   Merchandising   161

15   Strengthening the Relationship   169

16   Professional Education Development and Your Sales
         Career Ladder   175

References   185

Index   189

About the Authors   195

Figures and Tables

FIGURES

3.1	Intensive versus Selective Channel Strategy	29
4.1	Example Wine Sales Spreadsheet	45
5.1	Why Consumers Drink Wine	52
5.2	Consumer Wine Preferences	52
6.1	Be Committed to Your Customer's Success	68
9.1	Helping Your Customer with a Gap Analysis	105
9.2	Gap Analysis Identifying Missing Wines in a Customer's Portfolio That Are Selling Well in Similar Settings	107
10.1	Succeed by Helping Your Customer Succeed	110
12.1	Reassuring the Customer	138

TABLES

2.1	Three-Tier Pricing as a Percentage of Suggested Retail Price	18
3.1	Constellation Wine's Consumer Psychographics	25
10.1	Translating Features into Benefits	111
12.1	Three Basic Types of Customer Objection	139
13.1	Presenting Features as Benefits	155
13.2	Probing for an Opportunity That Can Be Supported	156
16.1	Wine Education and Degree Paths at US Colleges and Universities	179

Preface

This book represents the authors' understanding of the world of sales professionals within the global wine industry. It provides the reader a means of understanding the principles, strategies, and practices used by some of the leading wine producers, distributors, importers/exporters, specialty retailers, and their most effective salespeople. The foundation of this book is over six decades of combined research, consulting, and teaching in personal selling skills gleaned from countless interviews with effective sales professionals in the wine and broader hospitality industries. Many of their ideas have been incorporated into a framework called "consultative selling skills." Moreover, they have been validated through interviews with the buyers they serve as to what makes a salesperson effective. Collectively, what they tell us is that the skills required to be a successful salesperson can be learned and always improved.

Our motivation in writing this book is to better align our teaching and consulting with the goals of our students and clients. We are marketing researchers by training and began our respective academic careers teaching the marketing mix paradigm. Though the effort has been rewarding in terms of consulting and academic publications, we all independently began to realize that it did not serve our students particularly well after graduation. Marketing is currently taught and practiced from a retailing perspective: offering the right product, at the right price, at the right location (or method of distribution), with the right promotion to the right consumer market segment(s).

Wine is sold to consumers at the retail level as well as in the trade buyer-supplier environment, where wine producers sell to organizational buyers. These include wholesale distributors in the United States, importers/distributors in China, and big-box retail chains in Europe—who then, in turn, sell to retailers (or in the case of Europe, to the final retail or restaurant consumer).

The reality is that the salesperson in each of these situations is responsible for implementing the marketing strategies of his or her firm. Both in direct-to-consumer sales and business-to-business sales, it is the salesperson, supported by great products and effective marketing collateral, who is entrusted with a firm's most precious asset—its customers: consumers, organizational buyers, and channel partners.

Makers of fine wines know that there is a limited market for products, and they seek to attract loyal and profitable end consumers through their shared values and common interests. Similarly, great salespeople are selective when engaging with prospects. Research by VoloMetrex, a sales productivity firm, has demonstrated that top sellers build deeper relationships by dealing with fewer customers rather than by casting a wider net of shallower engagement. Savvy salespeople cultivate more profitable and sustainable accounts that inherently represent a good fit with their company instead of trying to close as many sales leads as possible. In the case of wine sales, great salespeople will study a potential customer's wine portfolio and consider where he or she can add value to the range before reaching out to the customer. In the same way, the retail salesperson will try to understand the consumer before suggesting an appropriate wine for purchase.

TO THE READERS OF THIS BOOK

For those readers who are also marketing oriented, the breadth and depth of career opportunities in wine sales and distribution are extensive and afford you the ability to put all your skills and knowledge to use. But we must be candid with you: it can be hard to get your foot in the door. This book should make it easier to obtain your first sales job.

These are some of the traits and skills an employer will look for in a sales candidate:

1. A passion for wine and an understanding of the complex structure and legal environment of the industry
2. The right previous work experience
3. Self-confidence, relaxed demeanor, polished communications skills, and ability to build rapport
4. Good chemistry—the candidate finds things in common with the interviewer

But even displaying all these traits in an interview does not ensure the sales candidate can deliver as needed. The smart director of sales knows the behaviors and skills needed and will structure the interview to assess them. One of our industry advisers asks every candidate to "sell me a replacement to the pencil I have been using for years." We will prepare you to respond to such a test in an efficient and effective way.

For those readers with previous wine distribution sales experience, this book will refresh your skill set and give you new concepts and strategies to add to your quiver—drawn from the best in the business. The role of the salesperson is changing, and the best have shifted away from chasing every potential customer to building lasting relationships with the best customers.

Acknowledgments

Many people were instrumental in giving freely of their time as we wrote this book. Chief among them are the following:

- Shawn Byrnes, retired, vice president of marketing, Ste. Michelle Wine Estates
- John Caudill, chef, marketing and sales manager, Sheridan Vineyards
- Doug Charles, owner, Compass Wine
- Vince Friend, CEO, Cordelina Wines
- Henri Gabriel, president and CEO, Advantage Distributing
- Chad Harding, general sales manager, Constellation Brand Northwest Region
- Curtis Mann, director of alcohol and beverage, Raley's Markets
- Brian Renney, general manager/EVP Oregon, Columbia Distributing

Any praiseworthy insights should be credited to these wine industry leaders who generously shared their knowledge and expertise. Any flaws in the book rest solely with the coauthors for not asking the right questions or listening more carefully to what each shared.

Section I

PROFESSIONAL FOUNDATIONS

In this section we explore and explain the basic foundations of the sales process and wine sales in general. The exchange of value is at the heart of any sale, but how you establish that value, and how you help your customer understand the value that you bring to the equation, is what can differentiate the average salesperson from the true master. Building on these principles will help you become a top salesperson in your company, your region, or your area of interest.

Chapter One

Wine Sales

Foundations of Success

Consider for a moment: the owner of a large wine distributor has just added a new wine label to the company's portfolio, or the restaurant manager of a fine-dining bistro adds a new wine to the list, or the purchasing director of a large grocery store chain has opened a new account with a wine distributor. Imagine, too, a consumer purchasing a unique wine style, region, or varietal. At some point, each of these customers was no more than a lead or prospect. In each case, someone in the sales department of the winery or the wine distributor did several things right—and in doing so, gained the attention and trust of the customer. He or she won the business.

There's an adage that nothing happens in the wine business until something is sold. That is true at every level of the sales network. In fact, another old saying in the wine industry is that "the wine isn't sold until the customer pulls the cork." And there are sometimes a lot of steps between winemaker and opening the bottle. As required by law, a significant proportion of sales in the wine industry come through business-to-business marketing, the niche of wine distribution sales professionals. Selling to organizational buyers in our industry requires more than just advertising that you are producing excellent wines and ready to begin taking orders. It takes a trained sales force to identify their company's best potential customers, develop a deep understanding of each customer's needs, engage customers by teaching them something new and valuable about the product and service, and navigate what is often a complex decision-making process to win the business all the way to that consumer who enjoys the wine.

Make no mistake: with more than 130,000 SKUs (stock keeping units) produced annually in the United States alone, the wine industry is a buyer's market, has been so for decades, and will remain so for the foreseeable future. Gaining face time with potential customers to give a traditional sales pitch is more challenging than ever. They will not want to be bothered by a potential supplier or salesperson who intends to sell them something they may not need.

This does not mean that buyers are not interested in new wines any more than consumers are not interested in new flavors or approaches. These potential buyers want to be in the driver's seat when it comes to meeting their own needs. They want to be sure a new wine can deliver. To create the level of *a*wareness, *i*nterest, *d*esire, and ultimately *a*ction (AIDA for short) on the part of buyers, it takes a professional salesperson to put in the up-front effort to probe for and accurately understand the buyer's needs and to provide solutions to those needs that are better than the competition's. This is true for the buyer of a large wine retailer just as it is for a consumer who wants to serve a nice bottle at dinner.

Selling to these customers requires the ability to build credibility and trust in you, the seller, and in the products and services you provide, because a lot can be riding on their purchase decisions. Credibility emerges as a function of expertise, reliability, and trustworthiness. Therefore, a fundamental trait that leads to sales success is the ability of the salesperson *to create and sustain trust* with his or her most important customers, whether they be local fine wine collectors or large organizational buyers.

WINE PRODUCTION AND DISTRIBUTION WORLDWIDE

According to the most recent estimates of the International Organization of Vine and Wine, 267 million hectoliters (mhl) of wine were produced for sale globally, of which 242 mhl or 90.6 percent were sold. One hectoliter equals 100 liters or 26.4 US gallons. Drinking those wines may be a pleasure, but selling wine in such a competitive global market is a serious business not for the faint of heart.

Italy is the world leader in wine production at 50.9 mhl, followed by France (43.5 mhl), Spain (39.3 mhl), the United States (23.9 mhl), and Australia (13.0 mhl). The United States is the country with the most substantial wine consumption at 31.8 mhl, followed by France (27.0 mhl), Italy (22.5 mhl), Germany (20.2 mhl), and China (17.3 mhl). Hence, some countries produce surpluses of wines for export, some import more than they consume, and some have a mix of imports and exports that roughly balances with their consumption.

Estimating the number of vineyards globally can be problematic, given that many winemakers are microbusinesses that do not have to be licensed to sell. In the United States, a winery has to be licensed and bonded before it can sell its wines. As an illustration of the complexity of the wine industry, there are 10,417 bonded wineries in the United States (4,285 in California alone), which, according to the Wine Institute (2016), are supplying 10.4 percent of worldwide wine production (US Treasury Department, 2016). Most of this US production is consumed domestically. In addition, 399 million cases were imported, representing 39 percent of all wine sold in the United States. This production is consumed by the estimated 40 percent of the US (adult) population who are wine drinkers, making the United States the largest retail market in the world (Halstead, 2016).

Adding further complexity to understanding the wine industry globally are the laws governing the sale and distribution of wines that are unique to each country. In the United States, wine retailers are required by federal and state law to source their wine through wholesale distributors, who in turn must be licensed by each state. According to the Wine and Spirits Wholesale of America Association (2017), there are over four thousand distributors, with sixty-five thousand full-time employees earning $5 billion in wages. These distributors range from relatively small boutique wholesalers to a few large national distributors who often control 80 to 90 percent of all wine distributed in each state. In Europe, wine producers can sell directly to wine retailers, which explains the lower retail prices in Europe. However, often wine exporters to Europe, as well as wine producers on the continent, employ contract sales companies who provide functions similar to US distributors but on a fee-for-service basis. Where needed, they assist importers/exporters with securing customs clearance facilities, navigating each country's unique regulatory systems, and developing and putting into action marketing strategies to penetrate and expand markets for their clients. Countries in Asia import far more wine than they produce, and wine importers also serve as wine distributors to their networks of retailers.

Size matters in the production, distribution, and retail of wine, with significant recent changes in each of these sectors (MarketLine, 2016). For instance, the number of wine wholesalers in the United States has decreased by 75 percent since the 1970s. Now, three distributors account for nearly 50 percent of the wine sales market: Southern Glazer's Wines and Spirits, Breakthru Beverage, and Young's Market Co. By their purchasing power, these three can command discounts that few small wineries can provide profitably. However, many small to midsize distributors thrive by providing their customers with both the wines they want and the support services that their clients and customers need to succeed.

Another dynamic force shaping the wine retail sector is the increasing propensity of consumers to purchase wines from grocery store chains, wholesale clubs, and big-box wine merchandisers. Though this trend is global, it is particularly evident in Europe, where according to the Netherlands Ministry of Foreign Affairs, a handful of organizational buyers control 80 percent of table wine consumption in western Europe. These large retailers exercise their purchasing power to their advantage on the price they are willing to pay and the support services required of their suppliers. However, a low price for a buyer does not inevitably mean high profits for any firm in the wine distribution chain. Providing a unique offering in terms of wine quality, style, and origin adds something distinctive to the buyer's portfolio—and is what consumers often seek. In addition, bundling the product with an ideal mix of support services is valuable. In such cases, buyers do not automatically require the lowest price because they can make a profit from having something unique to sell. A salesperson who understands this and who sells consultatively by helping buyers fill in the gaps and exploit trends in wine consumption will succeed.

Wine Spectator columnist Matt Kramer summed up the challenges facing the wine industry thus: "Today the problem isn't making fine wine. There is plenty of talent available. The problem now is selling fine wine . . . distribution (sales), you see, is the real problem." For many smaller wineries, this means selling the wine direct to the consumer and avoiding the distribution network entirely.

ORGANIZATION OF THIS BOOK

This book is designed to identify and close the gap between industry best practices and workforce training and preparation. Experience is an excellent teacher for salespeople, but to be useful, it needs a framework in which to assess performance. Often this is missing. Some salespeople have learned to sell exclusively through observing others, while some have learned by trial and error. Either way is harder than it needs to be. Unless the new salesperson knows what he or she is looking for, the mere observation of even a great salesperson will provide limited information. Frequently even the best salespeople themselves do not know or cannot express the reason for their success, which in turn makes it difficult to teach anyone else.

The framework of this book is designed to give you a realistic view of the knowledge and skills you need to master in order to be successful as a wine sales professional. The framework provides a roadmap for your future that you can start using right away and that has the depth to be helpful throughout your career. In today's harried business environment, salespeople need to grow and evolve professionally. Using a framework, like using a roadmap, is a more efficient and effective means to get where you want to go.

Throughout this book, you will read wide-ranging scenarios on customer needs and special circumstances throughout the wine distribution industry. On the one hand, as a book designed to teach a model of personal selling, it would be less complicated to limit the range of scenarios to a handful that would be integrated into each chapter. On the other hand, allowing for a broader variety of situations and illustrations provides a deeper appreciation for the challenges a wine sales representative experiences and may actually face across all distribution sectors. We chose the latter approach.

CHARACTERISTICS OF SUCCESSFUL SALESPEOPLE

Wine sales professionals are no different from sales professionals in other industries. All are charged with *initiating, developing, and expanding relationships with profitable customers*. To be successful, a review of what research tells us are the fundamental concepts of sales and characteristics of sales professionals that contribute to long-term success is in order.

One of the critical challenges all sales professionals face involves finding ways to earn the *trust* of the buyer. The trust formation process is a central construct in the personal selling literature. A lot is riding on an organizational buyer's purchase decisions. For example, a wine steward's job could be jeopardized if he builds a wine inventory that does not sell at the volume and price points that his restaurant owner expects. Trust is an expectation by the buyer that a seller will engage in the actions that will support not only the seller's interests but also the buyer's. The buyer initially assesses a supplier's trustworthiness through such perceived qualities as the salesperson's knowledge, reputation, commitment to service, honesty, listening abilities, communication skills, and helpfulness. It is later confirmed or rejected when the seller's organization delivers or does not deliver on what was promised. That's true whether it is a sales manager of a wine distributor calling on a restaurant or a tasting-room employee promising delicious wine to a consumer.

Success in wine sales also requires negotiated give and take. Most fine wines are limited production wines by definition. And most wine buyers are selective as to which labels they buy. A sales manager who allows a restaurant manager to cherry pick only from her best selections will quickly realize she will be disappointing her other restaurant buyers as well as their employers. A winery that sells out of its most popular wine, leaving only less attractive bottles in inventory, is going to struggle. How to strike the right balance in allocating inventory to meet the interests of all the relevant stakeholders is a crucial negotiated outcome the sales representative must manage.

Hence, the ability to initiate, develop, and expand a buyer's trust in a firm and its products and services is initially dependent on the relationship development skills of the salesperson. The salesperson also can think creatively and innovatively to find solutions for the customer that ultimately contribute to the success of both the buying and selling organizations. Understandably, great importance has been placed on identifying the characteristics that make a person a good candidate for a sales position in virtually all industries. A leading personnel testing company believes that one should look for a person who is stable, self-confident, goal-directed, intellectually curious, and good at networking. Others describe top salespeople as being organized, persistent, presentable, generally optimistic, socially intuitive, and incredibly honest. Research has shown that people are seldom born with all these traits. However, if you have the desire, attitude, and honesty (traits that cannot be taught), you can learn the rest over time through training, coaching, and sales experience.

THE REWARDS OF A SALES CAREER

Everyone sells every day. When you interact with customers (either internal or external ones) in any way, you are selling. Even if you're not looking for a business card with "sales" in the job title, you will often find yourself selling in some sense. In fact, the route to a position as a general manager of a winery or distributor often requires time spent in sales. Most firms understand that sales experience provides management trainees not only with an understanding of what it takes to create a customer but also the communication, persuasion, and interpersonal skills that will serve them well no matter what leadership position they later choose.

TWO BIG ADVANTAGES TO SALES CAREERS

But if you do choose to focus your career in sales, you will be rewarded. First, no other profession in the hospitality industry affords a person the ability to work primarily 8:00 to 5:00 Monday through Friday. There will be exceptions; in wine sales, when a restaurant unexpectedly runs short of a popular wine, you will want to ensure that it gets the needed delivery even if it means you are making the pickup and delivery outside of your typical work week. But compared to other hospitality industry jobs, that's unusual.

Second, wine sales reps are highly compensated individuals since they are responsible for the lifeblood of their employers (i.e., income). The primary goals of any compensation system are to enhance a firm's performance and to attract and motivate personnel. Generally, there are three basic methods of compensation: straight salary, straight commission, and combination plans of salary plus commissions or bonuses. Wine distributors vary between straight commissions (5 to 6 percent of sales) to a combination plan of salary and commission package. A combination plan generally contains a base salary with bonuses designed to encourage one of the following:

- Exceeding sales quota (in terms of dollar volume)
- Increasing sales of more profitable wine labels
- Increasing account penetration to existing accounts
- Making sales in low seasons and/or low-demand wines

Inbound call centers currently being developed by large national distributors may be salary only. By contrast, sales reps of relatively small distributors or wineries are often compensated on straight commission. This keeps the cost of sales for the employer in proportion to gross sales, is easy to understand and administer, and provides the maximum incentive to the most productive sales reps. This system is often used in winery tasting rooms, where employees are given incentives to meet sales goals.

Large wineries often start their sales reps on a combination plan involving a high base salary and low commissions. Over time, the sales director will reduce the salesperson's base salary and increase his or her commission rates to arrive at an ideal ratio that offers the salesperson income security and incentive.

GETTING STARTED IN YOUR SALES CAREER

Your first job in wine sales will likely bear little resemblance to your last. You may start out selling wine as a waiter in a restaurant or work the bar in a winery tasting room. One's first sales job for a distributor may begin as a merchandiser, managing the truck-to-shelf merchandising of a larger big-box wine store or grocery store chain. Next, it may evolve to a sales associate filling in where needed for the sales rep who manages the account or taking a management role on the floor of the tasting room or retail shop. The next rung may be a sales rep who services fifty to sixty accounts with annual sales quotas increasing 10 to 15 percent per year. Subsequent advances can also include becoming a portfolio manager of a distributor focused on making adjustments to the wine labels carried to achieve optimal sales and revenues.

A distributor's portfolio manager is the prospect, or customer, a winery's sales rep will target. A current trend in wine sales is for large wineries to embed a sales rep in a distributor. Often wine distributors will carry in inventory five thousand to eight thousand wine labels, of which the top one hundred labels contribute 85 percent of the distributor's annual gross sales. The winery's investment in an embedded sales rep is designed to educate, support, and incentivize the distributor's sales reps to increase sales for the winery and distributor themselves. This selling alliance between wine producers and distributors often leads to sales reps switching from sales positions between producers and distributors. The same is true of a top sommelier in a restaurant, who might be recruited to share his or her expertise and experience with a distributor's top restaurant accounts—and sell them wine at the same time.

The most senior sales professionals are often now fulfilling the role of key account managers, particularly in states that allow for franchise agreements. In such settings, the key account managers are responsible for alliances between buyers and sellers that promote mutually beneficial ties between two firms. In such *purchasing partnerships*, as they are often called, buyers receive quality wines and support services while sellers gain a significant portion of buyers' orders. The relationship enables both partners to plan requirements on a mutually beneficial schedule with mutually satisfactory pricing.

IN CONCLUSION

Wine sales and distribution is a dynamic and rewarding field that can stay exciting for a long career. Although we have discussed many traits considered desirable for success in sales, no one has discovered a specific set of personality traits that are absolutely necessary for success in selling. To put it another way, research has yet to find the profile of a "born salesperson." If you are interested in wine distribution sales, this should encourage you. Because of the wide diversity of selling positions and the adaptability of human beings, you may be able to "grow yourself" into an effective sales professional, even if you have only a minimum amount of knowledge and skill to begin with. In other words, if you are mature and motivated and are dedicated to improving your analytical and communication skills, you could become proficient in any sales. The process of becoming a good salesperson is dynamic, and the best professionals never stop getting better.

While no ideal set of characteristics has been found to guarantee success, several factors are strongly related to high performance: working hard and working smart. These two factors include the ability to set goals, put in the time and effort to learn your product and how it can benefit your customers, communicate well, and present yourself as a professional. To do so, you must also demonstrate maturity, dependability, honesty, and integrity. The good news is that these characteristics can be developed through thought and careful practice—and they are useful not only professionally but also personally.

DISCUSSION QUESTIONS

1. How is the wine business changing in the United States?
2. What role do "key influencers" like wine writers and wine experts play in your local market?
3. What are some of the skills that you think a top salesperson should have?

Chapter Two

Buyer-Supplier Relationships and Compliance Regulations in the Wine Industry

Sales professionals are found in all wineries, wine importers/exporters, wine distributors, and wine retailers regardless of size and setting. While some people mistakenly believe that success in the wine business is making excellent wine, we suggest that success requires both making good wine and selling it successfully to make a profit. If you can't do that, your enterprise will not be sustainable in any sense. The salesperson plays the pivotal role in winning new business. Sales are often described as the ability to create and sustain trust with profitable clients or prospects. The process can be thought of as *making and fulfilling promises.*

Every organization that is directly connected to the production and distribution of wine must have people dedicated to sales. Forget the old idea that if you build a better mousetrap the world will beat a path to your door. Offering quality wine with customer support services does not mean that sales will follow. Even in the best of conditions—when a firm has a good reputation and is receiving referrals through positive word of mouth—these are still only inquiries. It takes a person who is skilled in negotiating a sale to probe for the prospect's underlying needs and circumstances, show the prospect how his or her firm's features would translate into benefits, and then actually book the business at a price and terms that are profitable. A good salesperson can take a lukewarm prospect and turn it into an enthusiastic purchase of larger volumes. And the ultimate goal is to create a long-term relationship that continues to expand both in volume and frequency over time. This is true whether that salesperson is in the tasting room selling to consumers, calling on customers in retail shops or restaurants, or negotiating major deals with a chain-store buyer. While the scenarios may seem different, the same techniques will succeed in all these situations.

Selling requires far more skills than simply order taking. Even when the prospect initiates the contact, it is likely that he or she is contacting several other firms and weighing the options to identify the best solution or value. Think of a customer entering a wine shop. How many bottles will he or she buy? A good salesperson can change that number significantly. At the organizational buyer level, most wine business enterprises cannot afford the luxury of waiting for buyers to come looking for them. More often, firms realize that to compete, they must proactively go after new business and thereby create new demand for their wines. As part of the sales process, a good salesperson will do the following:

- Acquire in-depth knowledge of his or her wines and support services.
- Identify sources of leads or prospects within his or her territory.
- Prioritize specific leads that offer the highest quantity and quality of prospects.
- Study each prospect's existing portfolio of wines and suppliers, and determine where he or she can add value to the range.
- Determine the best method to make initial contact with each prospect.
- Stimulate the prospect's interest with an initial contact for purposes of scheduling a meeting, during which the prospect's needs can be further investigated or confirmed and the benefits of the supplier's wines and support service highlighted.
- Ask for the prospect's business.
- Cultivate that prospect to encourage repeat sales.
- Support and continue to develop the relationship.

These may seem like relatively simple steps, but they require a level of sophistication to be effective. More importantly, a successful salesperson will go beyond these steps. Because the consultative salesperson needs to provide both information and context to the customer, he or she is also in an ideal position to provide feedback to the winery by serving as the voice of the customer. By understanding the trends and needs of the market, the salesperson can provide invaluable information to the company's marketing team, allowing them to develop strategies in product development, marketing, and customer service that may be more successful at creating demand, enhancing the brand image, and driving sales.

TYPES OF BUYERS IN THE WINE INDUSTRY

Let's begin with the most apparent customer of all: the wine consumer. Every winery with a tasting room meets wine consumers every day, and the success of a small winery may depend on its ability to sell wine effectively to those consumers. In the following chapters, we outline the steps necessary to develop rewarding relationships with the consumers who walk into the tasting room, from learning what they want to responding to their needs and encouraging their enthusiasm. It's the same process for every type of buyer in the upstream distribution of wine, from the winery to the distributor, and from the distributor to the retailer. The only thing that changes is the input from the buyers themselves. But that is a crucial element to successful sales. One size does not fit all. Each sales approach should be tailored to the type of customer and the individual preferences of each individual customer—and this is equally true in the tasting room or when calling on the buyer of a major chain store.

Out in the retail marketplace, sales prospects are generally organized as on-premise and off-premise accounts. On-premise accounts typically include hotels, restaurants, bars, country clubs, casinos, and the like, where the wine sold is ultimately consumed on the client's premises. Off-premise accounts include grocery stores and specialty wine stores, where the wine is consumed at the customer's home.

In each case, the buyer will have a vision for what he or she wants and how your product might or might not fit into that vision. An Italian restaurant may want a wine list that is exclusively wines imported from Italy. A retail shop may focus on wines in specific price ranges and varietals based on its customer profile. Some wine specialty stores have greater incentives to be responsive to their customers with wine knowledge. When a buyer is buying for more than one location, such as a chain of restaurants or stores, the buying decisions often become more transparent. Buying decisions among chain grocery stores are generally managed at the corporate level, where the buyer is tasked with allocating shelf space that maximizes sales and profits for the retailer. Buyers often have a clear idea of what sells given the information available in their point-of-sales systems. For some buyers, wine knowledge may be less important than price and merchandising support. Understanding how to address these issues is discussed in detail in chapter 7.

COMPLIANCE AND REGULATIONS

It is important to note that very few sectors face the regulatory complications that we face in producing and selling alcoholic beverages. In the United States, sales are monitored and regulated, in one way or another, at the federal level through the US Department of the Treasury (Alcohol and Tobacco Tax and Trade Bureau), your local state alcoholic beverage control board, and even in the issuing and permitting of use licenses at the local level. The opportunities to make mistakes that could damage or even end your career are plentiful.

How did the United States become so overregulated? When the campaign to repeal Prohibition began in the early 1930s, those at the helm understood that to repeal a constitutional amendment would require the approval of 75 percent of the individual states. It's hard to imagine anything reaching that level of approval today. And so, to encourage each of the states to vote in favor of the repeal of Prohibition, the repeal campaign agreed to let each of the individual states regulate the sales of alcoholic beverages in their state. That's why the United States now has fifty different sets of regulations for wine sales in the country (not to mention that some states decided it was too difficult to convince their voters to approve a single solution for the whole state, so some states, like Texas and Kentucky, actually regulate alcohol sales by county).

What this means is that it would be well beyond the scope of this book to give you a full accounting of the appropriate regulations in all counties and jurisdictions. And given that these laws are also changing, we strongly recommend that you consult with a legal alcoholic beverage compliance attorney before taking any steps that might run you afoul of these laws.

During Prohibition, alcoholic beverages were distributed illegally and often via a system that included organized crime and vertical monopolies in many markets. To eliminate any possibility of this system continuing after Prohibition, the repeal included a complicated set of laws and regulations that created our three-tier system: producer, wholesaler/distributor, and retailer/restaurateur. The specific goal was to keep these levels completely separate and to eliminate commercial bribery or leverage. This means that producers cannot have an ownership stake in distributors, and distributors cannot own retailers or restaurants, nor can they give anything of real value to the lower tiers of the system.

The following issues are frequently involved in wine sales:

- You cannot require, demand, or attempt to coerce customers into purchasing any of your products. If they want to buy one of your wines, you cannot require them to purchase other wines to "qualify" or earn the right to buy a product that is in more demand.
- You cannot require or demand that they sell your wine at a specific price or maintain a price above or below a specific amount. Once they have purchased your wine, they own it and can sell it at whatever price they want.
- You cannot threaten any action, or refusal of goods or services, as a way of influencing their pricing decisions.

- In some states and regions in the United States, pricing is set by the regulatory government agency.

And this one is quite important:

- You cannot promise a customer anything, either goods or services, as a way of influencing their pricing decisions. What does that mean in real life?
  - You cannot provide a list of where your products are sold unless you include all the locations and outlets. That means that you can't suggest to consumers that your wines are available at XYZ store or Chez XYZ restaurant without mentioning all the other stores and restaurants where your wines are sold. Doing so is considered a real value to the store or restaurant and indicates preferential treatment. This also includes social media: focusing social media attention on a promotion that you organize at a specific store or restaurant is usually deemed a violation because it is seen as promoting that store.
  - In some states, sweepstakes promotions are legal, but the prize is limited to a value equal to one dollar. That's not much of an incentive. In other states, sweepstakes are expressly prohibited. And contests, where you give prizes based on skill or successful activities, are almost always illegal.
  - Free shipping is not something you can offer to consumers. There are ways to work around this with a competent attorney, but advertising free shipping is illegal, period.

Does all this sound complicated? It is. That's why you should make sure you have professional legal counsel familiar with your specific market and regulatory authorities before you tackle any of these issues.

PRICING CONSIDERATIONS

As noted above, you cannot legally attempt to influence the pricing of your products once you have sold them to a retailer or restaurateur, but it's worth spending a little time talking about the pricing of wines in a general way because it is a critical issue in wine sales. Let's start back at the three-tier system. Most producers will provide a suggested retail price (SRP) for their products. (Remember, the actual price is up to the individual account, but you can suggest a retail price based on the standard markups in the wine business.)

As a general rule, the first tier of producers sells its wines to the second tier at approximately 50 percent of that suggested retail price—and this pricing is usually given as the price per case throughout the system. If a bottle of wine has a suggested retail price of twenty dollars, the suggested retail case price is $240 for a standard twelve-bottle case of 750 mL bottles. That means the wines will be sold to wholesalers and distributors in the second tier at 50 percent of that amount, or $120 per case. This is an industry standard.

The second-tier wholesalers and distributors will then sell that same case at two-thirds (66.67 percent) of the retail price, which in our example works out to $160 per case. In terms of a percentage of the retail price, the distributors make a profit of 17 percent of the SRP. But regarding the purchase price they paid for the wine, it's a 33.33 percent profit.

The retailer purchases the wine from the distributor at 67 percent of SRP price and might sell it at the suggested price: $240. So concerning the suggested retail price, the retailer would make a profit of 33 percent, but in terms of the price it paid for the wine, it's a 50 percent profit margin.

Restaurants are a special case because they often mark up their wines far more than a retailer does. There is no industry standard markup for restaurants, although some have suggested something like a multiple of three or four times cost. But with wine-by-the-glass programs, featured wines on special, and other kinds of promotions and hand-selling opportunities, restaurants do fall into a special category that we'll talk about in more detail later.

Here is a chart to make this a little clearer. This time we are going to use a ten-dollar retail bottle price as an example (see table 2.1).

How much wiggle room is in this system? Let's begin at the top. As a producer, you have to sell the wine for more than it costs you to produce it, or you won't make a profit. That's obvious. But there is no standard profit margin or markup for wine producers. Some wineries invest a lot in marketing, public relations, and promotion, and can consequently charge far more

Table 2.1. Three-Tier Pricing as a Percentage of Suggested Retail Price (SRP)

Producer sells at 50% of SRP	$60/case
Wholesaler buys at 50% of SRP	$60/case
Profit	$20/case
Wholesaler sells at 67% of SRP	$80/case
Retail account buys at 67% of SRP	$80/case
Profit	$40/case
Retail account sells at 100% of SRP	$120/case

than their production costs for their wines because the wines are in high demand. Other wineries make a living by keeping costs low and fighting to be perceived as an excellent value for money. This is a decision that every winery has to make for itself as part of its business plan.

However, in the wine industry, there is a crucial element to pricing that cannot be ignored. Wine consumers strongly believe that more expensive wines are of higher quality, and thus the price a winery sets for its wines determines not only its profit margin but also to a large extent the consumer perception of the quality of its wines. With this in mind, it is obviously to every winery's advantage to keep prices as high as possible. After all, a high retail price is the surest way to convince consumers that the wine is of high quality. Except that producing an expensive wine that doesn't sell isn't a business model that will work for anyone. In the end, every winery has to find that balance and must price its products to sell effectively in the marketplace: high enough to convey the image of quality, low enough to attract enough consumers to buy it. And keep in mind that many sommeliers will tell you that if wine on their list doesn't sell well, they often *raise* the price to make it more attractive!

Once the wine has been sold to a distributor, the winery has lost all control over pricing. Distributors buy wine for only one reason: they want to sell it quickly and profitably. If the wine is not selling effectively in the market, for whatever reason, distributors are far less concerned with the image of the product in the marketplace and far more interested in making sure that they get back the money they spent buying the product.

On the one hand, a distributor might very well discount an unpopular wine to help it sell more quickly—and that will have precisely the opposite effect of what the winery was hoping to achieve. The wine is now on the market for far less than the suggested retail price, and consumers will view it accordingly. At the same time, the distributor is unlikely to reorder a product that it has had difficulty selling—leaving the winery with both a damaged quality image and fewer prospects for sales in the future. That's not what the winery was hoping to achieve when it priced its wines.

In this regard, wine retailers are very much like distributors. They want to carry products that sell quickly and effectively. They won't reorder slow-moving items and are even more willing to discount wine to get it off the shelves and out of the store.

For all these reasons, it's important for a winery and its sales team to be in tune with the market—to understand what kind of pricing model works and which ones might well fail. It's important to realize that reducing the price of a wine has a negative impact on its perception of quality. But it's also important to understand that when you ask distributors or retail accounts to sell more of your wine, the first thing they will mention is a reduction in price. By offering a short-term discount on a wine, you can sometimes attract more attention to it at every level of the three-tiered system, and this is a usual strategy for some producers. A one-month special discount, particularly when it is applied to large-volume orders, can help jump-start it through the system. In this case, distributors may not pass on the discounted price to their customers—preferring to keep the increased margin as profit but making sure that their sales staff give the wine an added boost of attention. Or they may pass on a portion of their savings to their customers and encourage them to keep an additional margin as profit.

Executed expertly, this can generate a lot of wine sales for a product, but if the product doesn't sell through the pipeline, the winery is faced with the same old problem: a lot of wine on the shelves that is not moving. And those discounts quickly find themselves working through the system all the way to the customer. Even if the discount plan works correctly, you may see that the winery has now laid out a strategy that cannot be escaped. Both the distributor and the retailer know that the winery can offer discounts and may well ignore the product until those discounts are offered again. If this is the case, the winery might have simply reduced prices to begin with and moved to a lower price category.

When it comes to on-premise sales, restaurants and their related businesses are a special case. Most restaurants include high profit margins in beverage sales as a key element of their business plan. They may not make a tremendous amount of money on their food, but they hope to make up for that in wine, beer, and liquor sales. Prices for wine in restaurants are always considerably higher than in retail shops. Selling wines to restaurants requires a careful study of each business so that you can create a pricing relationship that makes the most sense.

Many restaurants have become increasingly aware that their customers often compare their wine prices with prices for the same wines in local wine shops—and the advent of smartphones has undoubtedly made this easier. Many diners lose their thirst for a nice bottle of wine when they see it priced at 300 percent of its price in the supermarket next door. As a result, restaurants are often interested in products that may have limited distribution so that the supermarket is less likely to carry them. To sell these lesser-known wines, they train their staff to hand sell the wines at the table with stories and enthusiasm. Pricing of individual wines, in this case, is less of an issue for the restaurant, as long as the wines fit into the general pricing category of other wines on the list.

Other restaurants don't expect their servers to be wine experts and prefer to carry well-known wine brands that sell themselves. They will want to know what kind of profit margin they can negotiate if they promise to purchase a lot of wine over a long period of time. Others may carry a limited wine list but then offer a broader range of wines by the glass, where the price markup is less obvious but can be quite profitable. These restaurants will also want special pricing from the distributor in return for making its wine one of those highlighted in the by-the-glass program.

In the end, wine pricing is both an art and a science. Understanding how it affects both sales and consumer perception is critical to success. As we continue to explore the process of consultative selling, we'll address how pricing can play a role in building long-term relationships with your customers and clients—and how pricing is not always the answer to every wine sales challenge.

DISCUSSION QUESTIONS

1. How is your local market different from markets in other areas of the country?
2. If a wine in your portfolio sells out in six months, what do you think you should do?
3. Are all wine buyers motivated primarily by making a large and rapid profit?

Chapter Three

The Exchange of Value between Buyers and Sellers

The sales force is arguably the most highly empowered group of employees in most companies. Often working alone and relatively unsupervised, salespeople serve a unique role for both selling and buying organizations. A selling organization must rely on its sales force to generate the revenue needed to sustain and grow its organization. In a winery tasting room, particularly for a winery that sells exclusively through the tasting room, these salespeople hold the future of the winery in their hands. Buyers rely on the salesperson to ensure that once a sale is made, the selling organization will follow through on its commitment to advance the buyer's interest in this specific exchange. In other words, for salespeople to be successful, they must influence mutually beneficial outcomes for both their employer (the selling organization) and the buyer. If you want to sell more wine, to consumers or the trade, your customers have to be happy with what they are buying.

Ethics is particularly relevant to the sales profession, where the salesperson must navigate through self-interests to find win-win solutions. The importance of sales ethics is further strengthened by the view that sales ethics can help sales organizations generate new customers and maintain and strengthen relationships with existing customers. Our trade interviews support the view that salespeople who fail to act with the best interests of their customers in mind underperform those who do.

What is of value to the seller's organization are sales that generate profits. A winery's ability to generate sufficient sales at a profit ultimately determines whether it survives and thrives as a business entity. With that said, a winery may be willing to forgo short-term profits in a sale that allows it to penetrate new markets, demonstrate its capabilities to a new potentially high-valued customer, or generate needed cash flow during times of low demand.

Buyers, on the other hand, purchase value. But what is value? Almost every winery explains that its own wines represent "good value for the money." Fair enough, but any consumer who buys a bottle believes that the wine being purchased is a good value for the money. It if were not, the consumer would not buy it. Therefore, we need something a little more specific as a way of defining value in this chapter. Value can be defined in the following equation.

Value = Benefits Received − Costs

The perception of value to the buyer can be enhanced in one of four ways:

1. Lowering the price of the product
2. Reducing the hassle by making it easier for the customer to acquire and use the product
3. Improving the product benefits
4. Adding value through the consultative selling process

The perception of value can be enhanced by the salesperson who concentrates on the benefits received. The key here is for the salesperson to identify things that the buyer values that are available at low cost to the seller. Bar promotions, complimentary tastings and samplings, sponsorship of an event, and wine dinners are examples of relatively low-cost options a winery can provide that are valued by distributors. Not surprisingly, many of those same options are attractive to the winery's wine club consumers. In another example, a world-famous, top-rated wine from a leading wine region will provide more value than a top-rated wine from an unknown region by an unknown producer. The quality of the wine may be quite similar, but the value in the market is wildly different.

The perception of value can also be enhanced through the consultative sales process. Salespeople who gain a full understanding of a customer's needs and then use that information to communicate to the customer how their product or service can satisfy those specific needs also enhance the consumer's benefit equation. This notion is fundamental to the consultative sales process that is the heart of this text. The following quotes are from a director of sales, demonstrating the added value buyers gain from salespeople who do more consulting than selling:

- "Increasing knowledge of wine category management supports velocity and moves my trucks." —Customer
- "With limited time, I buy from a consultant, not a knock on the door from a here-today, gone-tomorrow wine rep." —Customer

We often hear consumers in the tasting room admitting that they have visited the winery or purchased more wine because they have a special relationship with one of the staff in the tasting room. In every case, it's the consultative relationship that makes the difference.

Constellation Brands, one of the largest wine companies in the world, shares that the paradox of choice of thousands of wine SKUs available to their buyers, which may allow them marginal financial gains from switching to lower-priced wine alternatives, is outweighed by the trust and confidence the buyer has in the salesperson and the organization he or she represents. They have found that there are a few recurring types of buyers. Below we distill the main types of buyers along with their key characteristics and an action plan for effective sales.

In a similar vein, Constellation Brands has an ongoing research project called Genome that helps identify wine consumers by their psychographic profiles. It's well worth a visit to the Constellation Brands US website to learn more about this project and its findings, but table 3.1 summarizes some of the results.

Table 3.1. Constellation Wine's Consumer Psychographics

Consumer Segment (% of consumers)	Consumer Profile
Price-driven (US: 21%, English Canada: 13%, Quebec: 11%)	I believe you can buy good wine without spending a lot. Price is a top consideration.
Everyday loyals (US: 20%, English Canada: 17%, Quebec: 13%)	Wine drinking is part of my regular routine. When I find a brand I like, I will stick with it.
Overwhelmed (US: 19%, English Canada: 21%, Quebec: 25%)	I drink wine, but it does not play an important role in my life. I don't enjoy shopping for wine and find it complex and overwhelming.
Image seekers (US: 18%, English Canada: 16%, Quebec: 11%)	How others perceive me is important. I want to live a life that impresses others. I want to make sure the wine I choose says the right thing about me.
Engaged newcomers (US: 12%, English Canada: 24%, Quebec: 25%)	I'm young and new to an intimidating category. Wine is a big part of the socializing I do. I'm interested in learning more.
Enthusiasts (US: 10%, English Canada: 9%, Quebec: 15%)	I love everything about the wine experience. I love researching purchases, reading reviews, shopping, discussing, drinking, and sharing with others.

Consumers in other markets may have different motivations or habits. In China, for example, red wine is much more popular than white because red is the color of success and celebration while white is the color of death. In Germany, a higher percentage of wine consumers actively seek discounted prices compared to other countries. And in many Mediterranean wine-producing countries, the preference is for a locally produced wine, even over wines produced in nearby regions of the same country. With this in mind, it is vitally important that good sales professionals learn and understand the motivations of their clients' consumers in their own market. Only then can they become successful consultants to them.

BEST ALTERNATIVE TO A NEGOTIATED AGREEMENT

Wine sales, particularly in the trades, involve the give and take that is the core of negotiated agreements. While some wines are so rare and in such high demand they sell themselves, most do not. In these cases, the chances of winning a new account, or maintaining an old one, can be enhanced where the salesperson is given a degree of latitude in responding to a buyer's concerns and, at times, in providing incentives to overcome the agent's resistance to buy. Though responding to a customer's objections is discussed in greater detail in chapter 12, at this juncture it is important to understand when to negotiate and when to walk away to ensure that beneficial outcomes from negotiations are achieved.

The exchange of offers and counteroffers—and the inherent, inevitable concessions—should be governed by one's plan B, or what Fisher, Ury, and Patton (1991) coined as one's BATNA. BATNA is short for "best alternative to a negotiated agreement." Salespeople should never agree to make a concession in a negotiation unless it exceeds the value of their firm's best available alternative or if there is a strategic value in doing so. For example, it would be unwise to offer a heavily discounted wine during high-demand seasons. In such a case, it would be appropriate to offer regrets to the buyer that discount prices are unavailable and risk losing the business. This may even induce a perception of rarity, which can add value in the world of wine. On the other hand, from a cost-benefit perspective, it may be worth offering a concession that gets your wines on the shelves of a large-volume wine retailer. However, buyers have plan Bs or BATNAs, too. Gauging their BATNAs as best you can should influence your decision about the necessity and extent of a concession.

Attempting to understand the buyer's BATNA, particularly when he or she is employing hardball tactics, provides you an advantage. Assuming in the case above that you know there are few alternatives in your vintage or varietal during the season in question, the low BATNA of the buyer gives you little reason to make concessions that reduce the value of the sale. On the other hand, if there is a range of equally attractive alternatives to your wines available to your buyers, you should not be surprised that they will use this form of power as leverage in demanding additional concessions from potential suppliers. When faced with a weak BATNA (negotiating with a buyer with a strong BATNA), the wise wine salesperson will focus on buyers' dominant requirements, offering incentives that will meet their needs in slow, measured intervals, asking for something in return.

Research suggests that sellers with an attractive BATNA should tell buyers about it if they expect to receive its full benefits. Explanations can be framed in terms that most people in business can understand, where during high-demand periods discounted prices are closed, leaving only the standard prices available. Alternatively, sellers with an unattractive BATNA should be careful about communicating it since it may weaken their position. Bluffing is never wise given that the other party may know more than you think they do about your competitive position, and they may call your bluff, leaving you with nothing but a tarnished reputation.

SALES CHANNELS: LARGE-VOLUME CUSTOMERS—CHAIN BUYERS

It's one thing to sell wine in your tasting room. It's another thing to sell it to a retail shop or a restaurant. And it's often quite a different process to sell wine to a larger customer like a chain of restaurants, hotels, or supermarkets or to large-volume customers like airlines or cruise-ship lines. The process can be much more complicated, not only because of the volumes involved but also because of the impact of your decision on profit margins, inventory, and the viability of your company. The choices about which sales channels should get your top wines, your best prices, and your strongest support are decisions that will affect every element of your winery.

How important are these kinds of channels to a winery or a salesperson? They can play a major role in a lot of ways, so it makes sense to look at the larger picture. Each channel offers a different mix of volume, price points, and brand enhancement possibilities, and every winery needs to develop both a short-term and long-term strategy toward channel management. The bottom line is that the use of a single sales channel will limit a firm's sales performance to whatever that particular channel is able to do well. Smart wineries look for ways to expand their sales channels while maintaining their brand image and pricing strategies. And each of these channel customers will have a different perception of value.

At one end of the spectrum, the Internet as a sales channel works well for customers where price is more important than wine selection, quality, and support services. In such a setting, the cost per sale can be reduced for sellers, allowing them to offer better price concessions to buyers while maintaining profitability. The bottom line is that the use of a single sales channel will limit a firm's sales performance to whatever that particular channel can do well. Wines that have achieved a better reputation for quality and are in higher demand should employ only those sales channels that support and enhance their reputations and profit margins.

On the other end of the spectrum are the hand-selling opportunities that offer customers both endorsement and product knowledge in addition to the wine itself.

For a smaller winery, direct-to-consumer sales allow you to sell your wines at full retail price, which is a significantly higher profit margin. And this process will enable you to create and manage the brand perception almost completely. It's a perfect solution for some wineries, but unless you make fewer than ten thousand cases and can sell everything you make through the tasting room, that is not an option for you. Instead, you'll need to use brokers and distributors to sell your wines on a larger regional or national basis.

On-premise wine sales have often been used to help launch or develop a brand. Restaurants, particularly smaller ones, are likely to be more receptive to a new product that offers their customers a unique experience, flavor, or region. And these smaller restaurants can use their well-trained service staff to hand sell the bottles to customers. (Smaller retail wine shops can do this as well.) It's not a great way to achieve high volume, but such a sales effort will introduce a new product into the market. When you are starting out, that's the name of the game.

Off-premise sales tend to focus less on developing demand for a new product and more on generating volume sales to increase revenues or deplete inventory. Of course, there is a wide range of off-premise sales accounts. Some smaller fine wine shops offer much the same benefits as on-premise sales, with the bottles being hand sold to interested consumers. Supermarkets and discount houses are more likely to focus on price rather than brand building but can also move a much larger quantity of product. While each bottle of wine sold to a restaurant will find its way into the mouths and consciousness of a table full of diners, a case sold to a retail shop might be purchased by a single customer who carefully logs it into his cellar—not to be opened for another five years. That won't help build awareness for your new brand much at all.

In the end, no single channel can do everything well or competitively.

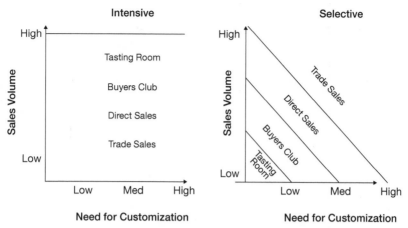

Figure 3.1. Intensive versus Selective Channel Strategy

Channel mix is a concept used by marketers to describe the multiple sales strategies they use to *go to market*. The purpose of a channel mix is to intercept a broad range of sales transactions in order to increase sales volume and market share. Channel mixes come in three forms: intensive, selective, and hybrid. Intensive channel mixes provide overlapping coverage of a market or markets where channels compete with each other. Selective channel mixes assign an individual channel to one specific market without overlap or integration with the other channels.

The foundation of channel strategy is a firm's understanding of each market segment's consumer behavior. The channels added to the mix should be those that collectively reach and profitably capture one's share of the transactions in each market segment.

So why not take a spaghetti-on-the-wall strategy, throw all of your products out on all the sales channels available to you, and see what happens? If you do this, you are chasing sales over profits. And you will have no control over the brand itself. The problem with such an approach is that more channels cost more money and inevitably lead to channel conflicts. Long-term profitability should always be considered in developing a sales channel mix. In designing a mix of sales channels, consider the following:

1. Multiple sales channels may end up chasing the same sales. Coverage of a market by multiple sales channels may create channel *shift* (customers moving from one channel to another) without channel *lift* (new sales). Your sales channels will just be cannibalizing each other. Though it is

difficult to estimate, one should assess how much new sales volume will be produced by adding a new channel to the product-market mix. If the answer is "not a whole lot," it may make sense to defer channel expansion to hold the cost of sales down. If you are selling your wines in the little wine shop just a block from your tasting room, the conflicts can become quickly apparent. If the shop discounts your wine, will you follow suit? Or will you keep your prices at 100 percent of SRP—which will encourage customers to buy from the shop? If you undercut the wine shop on price, you can expect it to discontinue your product pretty quickly. There is no winner there.

2. Not all channels will be profitable in every product market. Each channel should be focused on the product-market mix that can conceivably generate profits. Direct-to-consumer sales usually require staffing a tasting room, and that adds operation costs. But selling all of your inventory to a huge chain or box store might put far too many of your eggs in one basket and would likely be the result of heavy negotiations of a price. You'll sell your wine, but not at the price you wanted. And that can negatively affect the brand image for a long time to come.

3. Fine-dining restaurants generally avoid carrying wines that can be purchased at the local supermarket because it is too easy for their customers to compare prices and complain about the markup in the restaurant. If your wines are currently carried in a few good local restaurants, you will have to seriously consider the benefits (higher sales volume) against the drawbacks (losing restaurant accounts and brand image) when you offer those same wines to the local discount market chain.

   Take, as another example, a distributor's attempt to create flexibility in how reorders are taken from its client base of small restaurant customers. Though it may be profitable for a salesperson to personally take the order from the restaurant manager of a four-thousand-room hotel in Las Vegas, imposing the same model on a small restaurant that purchases two hundred to three hundred dollars' worth of product twice a week would be profitable for neither the distributor nor the salesperson who is working on commissions. Instead, the distributor focuses its field salespeople on securing the first business from these small accounts and hopefully migrating them to placing their orders over the Internet or by fax.

4. Manage the conflict where possible between channel members. In a product market covered by multiple channels, competition between channel members for sales is inevitable and needs to be managed. Channel conflict is just another way of saying that customers are provided with multiple purchasing opportunities by the same service provider. A national sales team may "steal" a sale from the local distributor (an example of channel

shift), yielding no direct gain for the winery. Such competition can be a good thing in that it keeps all channel members on their toes. However, it can be confusing for customers when they are provided different offers for the same product, often at different prices—and it can alienate the distributor in a big way. A solution many firms find more advantageous is to encourage their competing direct channels to work together (a topic we will address next in this chapter). On the other hand, conflict between a firm's direct channels and its indirect channel partners is a whole different story.

A winery that sells wines in a tasting room in the same market as its retail, distributor, or wholesale partners usually does so at a risk of voiding the supplier-distributor relationship. For this reason, many wineries choose to make smaller lots of wines that are for sale only in the tasting room, thus avoiding the conflict for at least these wines.

Thus, for a small to medium-sized winery, you might find a hybrid strategy that works, including the following:

- Offer a selection of limited-release wines that are only available at the winery to maximize tasting-room sales, wine club sales, and visitor center traffic.
- Create a line of reserve wines that are focused solely on the fine-dining trade, with a goal of increasing the visibility of the brand in attractive luxury settings. These sales may not generate high profits, but they certainly add cachet to the brand and help drive traffic to the tasting room.
- As the winery grows, release perhaps two or three primary varietal wines into general distribution for sale in specialty wine chains, supermarkets, and other large markets.
- Avoid selling wines to large discount houses or airlines in a hybrid strategy until the winery is prepared to produce a wine that can be sold to this segment profitably. When a winery employs this strategy, often it does so under a slightly different label to avoid brand confusion with its existing consumers.

For some wineries, pursuing sales directly to large-volume customers like national chains and airlines falls outside of their work with regional distributors. But that can also alienate those distributors, particularly in markets where the national chains have their purchasing offices. Such sales take potential profits out of the hands of your distributors, and that's not likely to make them happy. In this case, it is not "buyer beware" but "seller beware." To sell to those chain buyers, you will need to pass the wines through the local distributor there and pay that commission. And you need to include that in your pricing offer to the national chain.

So how are these large-volume customers different? To paraphrase Ernest Hemingway, they have more money, and they buy more wine. (F. Scott Fitzgerald once remarked to Hemingway that the rich are different from you and me. "Yes," replied Hemingway. "They have more money.") They may have a different sense of what value means.

These larger accounts will almost always have a much more defined sales process, from beginning to end. While you might be able to drop in on a retailer or restaurateur and hope to chat with him or her about your products, that is extremely unlikely to succeed with chain accounts. These accounts would be inundated with sales calls every day if they didn't implement some kind of screening and purchasing process, and your first step in selling to them is to understand the process. Following are a few examples pulled from real life.

A major airline uses a top-level master sommelier to select its wines. But before that master sommelier can taste your wine, you will need to go through a complex screening process. You will be asked to provide information including financial statements from your winery, references from other customers who have purchased significant amounts of wine from you, letters of support from your distributors in the relevant markets, and a signed agreement that will specify payment and delivery terms to the airline.

Once you have completed that paperwork, you might qualify as a certified vendor for that airline. At that point, you will be invited to submit samples of your wine that meet the criteria of the buyer's purchase request. Buyers might send out a notice to all their certified vendors that they are looking for 3,500 cases of chardonnay at a case price of sixty-eight dollars. If you have a wine that fits that criteria, you can then send samples. At that point, the sommelier gets involved. He tastes through all the submitted samples and selects the ones he likes the best. Then his selections are turned over to a purchasing manager at the airline who negotiates the final deal.

At a major wine retailer, a top wine expert tastes through all kinds of wines on a daily basis. She is happy to meet with salespeople, or at least accept their samples, which she then tastes and rates based on her palate and her appreciation of the wines. But she doesn't make any purchasing decisions. She may give your wine a rating of ninety-four points, but that doesn't mean the store will buy it. Instead, a purchasing team will then decide how the wine fits into the overall strategy of the store, and they will also estimate the relative value of your wine compared to other wines in the category based on their price. While your wine may be delicious, they may decide that they don't need a ninety-five-point white zinfandel that sells for thirty-four dollars a bottle on the shelf.

If you notice a pattern here, good. Companies that buy large volumes of wine want to make very sure that their purchases are thoughtful and effective. They want to avoid an "impulse buy" that might leave them with thousands of cases of wines that won't sell in their store or won't work in their airline. And so they make sure that more than one person is involved in the decision-making process, and they are more cautious concerning the kinds of companies and wineries they do business with.

The point is, multiple channels in a market are independent entities that sell against each other. In a growing market, this may not be a problem since all channel members have room to grow. However, if every sale comes at the expense of another channel member, something has to give: either fewer channel partners or fewer direct sales channels. Any realignment of channels should be focused on customer preferences and how they wish to buy and the long-term goals of the winery and its brand aspirations.

CHANNEL INTEGRATION

It is smart to employ different channels to achieve different goals for the winery and in each stage of the sales process. As we will explain in following chapters, the sales process can be thought of as a series of discrete tasks that involve prospecting, qualifying accounts, exploring needs and proposing, closing, and providing after-sales service and support.

Is it necessary to have one channel perform all these tasks? A more profitable and more effective solution would be to create a division of labor within the sales process or, in other words, assign different functions to each sales channel. For instance, few wineries sell a lot of wine on the Internet. In general, online sales are only a small part of the larger volume sold to consumers in the tasting room. However, the Internet in combination with a call center may be an effective way to generate traffic for the tasting room and may also provide a kind of sounding board full of information for your regional or national sales team.

While distributors may be necessary to fulfill orders to a major chain buyer, it's unlikely that a distributor salesperson with five thousand wines in his or her portfolio will do an effective job at working your winery and its wines through the complicated sales certification process you'll need for success. Many wineries have either a broker or a specialized salesperson just for these national accounts.

Channel management becomes even more important when you are good enough, or lucky enough, to have a wine that gains national or international acclaim and demand. Who will get those cases? The most profitable answer is to sell them only at the tasting room. But how does that affect the top restaurants that have carried your wine for years before it was "discovered"? And those small retailers have also earned a piece of the pie if you want them to carry the other wines you make. This is a serious decision that requires a great deal of strategic thinking to solve.

Unless a winery can sell all its inventory direct to consumers through the tasting room or wine club, it will need to develop a channel strategy for its wines. Limiting sales to any one channel is either too limited in market reach and growth or too expensive. A hybrid channel model is often a better solution. Well-designed hybrid models can provide impressive gains in efficiency and profits. However, they add complexity to the sales process and need to be managed as systems.

The best use of a highly trained and compensated sales professional will be to cover large, high-valued, complex accounts that require excellent customer support. Simple transactions will be pushed down to less expensive channels where it makes financial sense to do so. But those simple transactions can build both volume and loyalty over time.

We must emphasize that account ownership is mandatory across all selling tasks for an integrated channel system to work effectively. In the case of major accounts, a single sales manager (or sales team) should have ownership of an account from the first sales call to the closing of the first account and the continuing of the relationship. For smaller accounts, the sales manager may be in competition with other channel members in getting the sale. In another hybrid model, the sales manager's efforts may be integrated with other channels to handle only those steps in the sales process where his or her skills are truly needed. In the team-based hybrid model, the account is not *owned* by one salesperson. Instead, the account is *owned* by the organization and *worked* by the team of professionals who create and manage it.

THREE TYPES OF SELLERS

Transactional

Transactional sales is the oldest form of selling. These salespeople sell simple and low- to medium-priced products and services. These salespeople sell in "single calls" in a single appointment with the buyer. Long-term relationships with customers are a costly luxury since the salesperson relies on many sales transactions to achieve revenue goals.

Today, many products and services are becoming commoditized. Simultaneously, buyers are becoming more sophisticated and have easy access to product and service specifications. Since these buyers know what they want and where to get it and have access to many product suppliers, they are primarily interested in one thing: *price*. What value can the salesperson perform in these buyers' "buying process"? The answer: *not much!* In fact, the salesperson in transactional sales incurs a cost that raises the price. So a competitor who sells without salespeople (over the Internet, for example) has a lower price.

In a transactional sale, the customer is interested in value and will choose the supplier offering the lowest price. An example of this might be new car sales. These days, buyers don't need salespeople to tell them the features and benefits of a new car; these can easily be found on the Internet. Salespeople are not required to negotiate prices; many dealerships have listed and non-negotiable prices. Also, buyers shop the Internet to learn invoice costs and then shop further for the lowest prices. The same thing has happened in limited-service hotels that have small meeting rooms and seek to provide space for simple meetings and events. Online request-for-proposal systems today are making rapid inroads into this competitive space—selling their rooms and meeting spaces as a commodity where the lowest price wins the business.

What will happen to today's salespeople selling in transactional situations? They will be transformed into Internet order takers or marketers of e-commerce wine to backfill stock par levels. Sophisticated transactional customers don't need a salesperson. And they don't build the brand of a winery, either. They simply offer the product for sale at a low price.

Consultative

Consultative sales is the second oldest of the three basic forms. It theoretically evolved by the mid-1980s and only today is being put into full practice. These salespeople sell high-priced and complex products and services. They must build long-term relationships with customers. They sell in "multicalls" over a long time period and usually need to influence "buying committees." Fortunately, many accounts in the wine business fall into this category of complex products and services. Wine is complicated, and a good salesperson can be a valuable ally to the buyer.

Here salespeople add value through all stages of the "buying process." Their primary function is to build relationships with primary buyers—from wine buyers and sommeliers to the purchasing committee of a corporate account—and learn how each company works through the sales process.

What will happen to today's salespeople selling in consultative situations? They will become more valuable to the selling company in reaching and serving complex profitable market segments. The company will invest more resources in these people. Pay will rise, and they will be increasingly supported by technology and continuous training and development. Employers will support these "lead" consultative salespeople with staff, and this staff will simultaneously be in training to learn and eventually become consultative salespeople. This career niche should have a strong future.

Alliance

Alliance sales is the newest and most cutting edge of the three selling modes. As the newest form, it is also the rarest. In its most complete form, it is hard to see a distinction between the seller and the buyer. In fact, the selling company forms a selling team that exactly matches the buying team. Teams are represented by cross-functional areas; for example, both teams include a technology person and/or financial person as the situation dictates. These teams are as permanent as the partnership. These strategic alliances are characterized by such qualities as the buyer and seller sharing information and expertise as well as physical resources such as a warehouse or computer purchasing system. The seller may even have a permanent office located at the buyer's facility. The "vendor/buyer" distinction found in both the transactional sale and the consultative sale is blurred in the alliance sale.

Wineries like Kendall Jackson, E. J. Gallo, and Chateau Ste. Michelle work together with distributors such as Southern Glazer's Wine and Spirits, Breakthru Beverages, and Young's Market to market and sell their wines into the marketplace.

What will happen to today's salespeople in light of the growing "partnering" of selling situations? It's logical to expect that "consultative" salespeople will be the initiators of new partnering arrangements. While "partnering sales" requires a strategic commitment from each organization's top executives, consultative salespeople who currently have solid relationships with major accounts should be able to perform a key "brokerage" role for these strategic alliances. In addition, these salespeople will retain an important long-term role in the partnering teams. This is indeed a career goal to work toward.

IN CONCLUSION

Salespeople play a critical role not only for their employer but also for their customers. You must understand what value means to the customer, and you must understand how the different sales channels will affect the ongoing success of the winery and its brands. To be successful over the long term, salespeople must influence and generate mutually beneficial outcomes for both their employer and the buyer. You and your company will succeed by helping your customer succeed. Knowing your BATNA and that of your customer can assist in determining when to offer concessions, when to hold firm, and when to walk away. Your BATNA is the only protection you have from playing a win-lose game.

At every level, you will encounter a wide range of people in your customer base. Your ability to understand what they want and offer it to them in terms that are not only acceptable but also beneficial to your winery will determine your success.

On a career level, your ability to develop such value-oriented relationships with your customers will be an enormous asset. Every current customer becomes a potential referral for future business. And every satisfied customer becomes a source of new business as he or she makes subsequent transitions into a new position or a new company. That smart young shop assistant who loves how you talk about the wine might someday become the buyer for a major chain. And the more you cultivate those relationships, the more successful you will be.

DISCUSSION QUESTIONS

1. What does the word *value* mean to you?
2. What items have you purchased that you think are good value? Why?
3. What items would you purchase even if you could not get a great deal on the purchase price? Why?

Chapter Four

The Organization of a Sales Force

Various sectors of the wine industry deploy their sales forces in unique ways. Some sectors are relatively simple, like direct to consumer, where a tasting-room manager or wine club manager is responsible for a single salesperson or small team handling all retail sales. A restaurant may have a dedicated sommelier or merely ask all waitstaff to sell wine as part of their job. Other sales forces in the wine industry trades are more complex, with sales managers assigned a geographic territory or specific market segment and tasked with managing and growing their existing accounts of organizational buyers as well as prospecting new ones. Wine wholesalers and distributors typically populate this spectrum, both involving a number of overlapping sales channels to reach and serve their targeted markets. However, all wine wholesalers and distributors organize their sales forces around two central tasks: (1) supplying wine for on-premise or off-premise wine consumption and/or (2) account management and territory management. Selling direct to consumers is the focus of the next chapter.

ACCOUNT MANAGEMENT

As previously mentioned, a sales professional is charged with initiating, developing, and expanding relationships with profitable customers. That's a perfect description of a wine club manager at a winery as well as a distributor's sales representative in a major market. Consistent with the consultative selling theme, a significant proportion of this charge involves working with existing accounts. Hence the term *sales manager* is a more accurate description of the task, given that the salesperson is actively managing accounts. A sales manager's effectiveness in this arena is measured by the degree to which he or she has built a productive, worthwhile, and enduring relationship with the customer. Trust, commitment, cooperation, conflict resolution, and information sharing have all been shown to have a significant impact on sales.

Beginning on day one, a sales manager is assigned either a geographic territory or a specific market segment to develop. Within this territory or segment, you will have existing accounts made up of businesses that have traditionally repurchased with your company. Your initial task is to familiarize yourself with each account and organize them into an appropriate contact management system.

In recent years, sales management analytics have increasingly moved away from the view that all customer accounts should be treated equally, accepting the reality that customers vary in terms of the value they provide (and can potentially provide) to the selling organization. Winery wine clubs often have different levels or identify certain consumers as deserving more attention and care—because they buy more and better wine. After carefully reviewing each account, the salesperson sorts each into one of several groupings, such as the following:

- A accounts: established accounts that generate a high level of business for your winery, wine distributor, or wine retailer. These accounts warrant the greatest attention in the form of staying abreast of the clients' full range of business needs, monitoring how well your company is meeting their needs and expectations, and generally trying to strengthen and expand the business relationship. Through personal sales calls and telephone and e-mail communications, you stay in frequent contact with these clients.
- B accounts: high-potential accounts. They may be established accounts that are already providing a reasonable level of business but have the potential to offer more (e.g., by you winning other pieces of their business away from competitors). These may warrant your attention at the same frequencies as A accounts, perhaps even more.
- C accounts: potential new accounts or established accounts with average potential. Staying in contact is important but scheduled less frequently.
- D accounts: potential new or established accounts with low potential. Follow-up on these accounts comes last, after all other accounts have been handled.

Your purpose in organizing your accounts in this way is to allocate your time so that you meet your sales goals. Again, your A accounts should receive the most attention given that they provide a recurring stream of business that you will want to preserve and, where possible, expand. B accounts are deserving of your attention as well, since all sales organizations at times fail their customers, thus providing the astute sales manager an opportunity when the client is open to alternatives. In addition, your B accounts can be influenced to switch suppliers if you can offer them a compelling value proposition to do so.

TERRITORY MANAGEMENT

Of the four Ps of classic marketing theory (product, price, placement, and promotion), the one where sales most directly plays a role is in the area of placement. In fact, it is a frequent issue of contention between the sales and marketing team, since the former is focused on meeting its sales volume numbers and the latter is focused on making sure the product is seen in the kind of venues that enhance the brand.

Many wineries see five distinct channels or categories of placement for their products: direct to consumer, on-premise, off-premise, chain accounts, and national distribution. Let's take a look at each one and evaluate how it affects both sales volume and brand enhancement.

The most obvious of all sales channels is the direct-to-consumer (DTC) outlet via a tasting room at the winery or wine shipments to customers or wine club members. This is clearly the highest-revenue channel, as the wine is usually sold at or near the full retail price, without the usual discounts that apply to sales to the trade. And the volume can be quite high. The authors know of at least one winery in the Napa Valley and another in Spain that sell approximately one hundred thousand cases of wine direct to consumers at full retail price. With that sort of volume, the winery is certainly making a solid profit. And many wineries in the United States sell upward of five thousand cases of wine DTC.

For smaller wineries, this is the first avenue of sales and should be the primary focus of all sales activities. Only when the winery produces more than it can sell through the tasting room should it consider looking at distribution to the trade, with one notable exception. And given the volume we have already mentioned for many wineries, the winery's first priority must be to build its DTC program to reach at least five thousand cases.

The one exception to this channel for smaller wineries might be a limited sales effort to top-level restaurants. This is known in the industry as the on-premise trade because the wines are consumed on the premises where they are purchased. On-premise sales can be effective at achieving two separate but essential goals. By placing the wines in high-quality restaurants, the winery can help build the perception of the brand as exclusive and high quality. And the patrons of the restaurant can also become fans of the wine, purchasing additional bottles through the winery's DTC program. It's good marketing, and it adds to the sales team's bottom line.

If a small winery begins to think about sales to the trade to augment its existing DTC program, a targeted on-premise campaign is the place to start. Restaurant sales have an additional value to wineries that are hoping to build awareness and increase the visibility of the brand. While a retail shop might sell a case of your wine to a single individual customer who could then put the wine in his or her cellar and age it for years, every bottle sold by the restaurant will be poured for two, four, or even more customers. Thus the consumer reach of a case sold to a restaurant is far higher than that of a case sold to a retail shop. Some smaller wineries even consider these sales a marketing expense, since they don't generate the same kinds of profit as DTC sales, but they do drive awareness and higher DTC sales.

Of course, as wine production continues to increase to meet the long-term business plan, eventually the winery will have to look at selling wine into the more extensive distribution network, reaching beyond the tasting room and a few top restaurants. And that brings us to a critical discussion of how placement affects both sales and marketing.

Small fine-wine shops will have very much the same effect on your brand as a top-quality restaurant and will often help build the brand by hand selling your bottles to the consumer. It's vital that you give them the stories and tools they need to do this successfully. But while a few fine-wine shops and a few top-level restaurants can add some cases to your bottom line and help build your brand, if you really need to reach higher sales volumes, you will have to aim at a larger game; a few fine-wine shops and a short list of restaurants won't do the job.

Chain stores and supermarkets can easily sell a thousand cases of a wine that is floor-stacked for the holiday season, although this is far easier said than done. Many wineries and distributors will be working hard to earn those few key placements, and the pricing and promotional activities that you will need to compete may be more than a small winery can manage. The same is true of a special promotion through a chain of restaurants via a by-the-glass program or another creative marketing effort. Bear in mind that these programs need to be set many months in advance, as holiday promotions are often sold into the larger chains by the middle of the summer.

But even if you don't generate an important promotion or holiday program, selling your wine into a larger chain is still a very effective way to move cases. The math is simple. If a supermarket chain has fifty stores and sells even one case of your wine per month per store, you will be selling twelve pallets of wine a year to that account. That's significant, and the stores will certainly ask for and expect some kind of pricing discount in return. The same is true, to a lesser degree, of chain restaurants.

But be careful how you proceed here. The local fine-dining restaurant that has carried your wine for years as one of its local favorites will not be pleased to see the same wine in a local grocery store for a much lower price. The restaurant customers will quickly begin to see that the restaurant is marking up its wines a lot (they all do), and the restaurant will quickly replace your wine with one where the pricing comparison is not quite so obvious and easy for its customers. While you will be selling more wine to the trade, your image may suffer from the perception that your prices are lower (wine drinkers always associate price with quality), and your wines may no longer be served at the best restaurants in town. Sales will be up, but your brand image will have taken a step down. As that situation continues, it will negatively affect your profits over the long term.

This frequently becomes even more tangible when you sell a large amount of wine to a major wine warehouse or big-box store that offers it at an even lower price. Many consumers have learned to check their phones for information about wine, and one of the first things they check is the price. If your wine is widely available through a large discounter at a low price, it will be very hard for any of your other trade accounts to sell a lot of your wine. The result again is even more volume but very little hope of being considered a highly prestigious brand.

Some wineries choose to offer no discounts in the hope of maintaining a consistent pricing strategy throughout the country. But once your wine is sold to a distributor, you cannot control what the distributor might do to achieve its own sales goals.

A winery with a significant inventory surplus is in a particularly vulnerable position. Facing the need to sell far more than its usual sales volume, the winery often chooses to offer massive discounts for a short period of time. The discounts may only last for weeks or months, but the impact on the winery's image with consumers can last for years. For this reason, well-managed sales teams often look to sell this excess inventory into a market with little contact with the winery's primary customers. A distant state, an airline that provides no pricing with the wines it serves, cruise lines, military PX stores, and the like all offer at least some hope that the wine might sell through this channel without affecting the larger brand image of the winery.

As salespeople, we are often asked to reach specific volume goals, but many times that challenge is complicated by the insistence that we not offer large discounts or sell to larger clients who might reasonably expect those discounts. Making sure that the winery management team is clear on both the objectives and the effects of these channels on sales will be critical to your future with the company and in the wine industry in general.

SALES CALL REPORT

Every time you meet with a customer, you will have all the information needed to create a sales call report. Capturing that information while it's fresh in your mind will assist in planning your next visit. When documenting your call, ensure that you do the following:

- Capture all notes from the call, including information gathered during the meeting.
- Be sure to write down any follow-up action items for the account.
- Record information through the use of your CRM (customer relationship management) sales technology.
- Update your preplan or needed documents to reflect success.
- Note any follow-up needed by the buyer.
- Determine who will receive the product.
- Record any changes to the account (buyer).

This is a good list to follow for both salespeople in the field and tasting-room employees selling wines to tourists. Using a good CRM program is critical for success. Add a date for follow-up and next steps for the commitments made during the presentation. This is one of the most important things successful salespeople do, and it also continues the sales planning process, as we have now planned our next visit.

Next, set a SMART objective next time you call on the account. Using the SMART acronym can help ensure a clear understanding of goals during your next customer call. A SMART objective is *s*pecific, *m*easurable, *a*chievable, *r*elevant, and *t*ime-bound.

EXERCISE

Now it's your turn to create a SMART goal, answering the following questions to be sure that you are following the SMART approach.

Your SMART goal example:

What is specific about the goal?
Is the goal measurable? (How will you know the goal has been achieved?)
Is the goal achievable?
Is the goal relevant to performance expectations or professional development?
Is the goal time-bound? (How often will this task be done? Or, by when will this goal be accomplished?)

SALES CALL TECHNOLOGY

Writing and transmitting sales orders throughout the day will help avoid costly mistakes. After leaving an account, process all orders so you can avoid the late-day rush. When writing your order, always verify that the information on the screen is correct before transmitting; errors and processing credits are very costly, and out-of-stocks are expensive. To transmit your orders accurately as a sales manager, you must be proficient in the use of technology.

In conjunction with your SMART goals, you should develop a communication plan that lays out how frequently and by what means you will stay in contact through technology. Over time you will determine how and when

Figure 4.1. Example Wine Sales Spreadsheet

	Wine	Date	Account	Buyer	Notes	SMART Goal	Follow-Up
OFF-PREMISE							
ON-PREMISE							

each client prefers being contacted and adjust your plans accordingly. Use of your contact management software can help you stay organized. In addition, your clients will vary in terms of how frequently they would like to hear from you. On the one hand, we hear from distributors that wineries are often too demanding of their time, to the point that they quit taking their telephone calls. Caller ID makes this easy to do. On the other hand, restaurant owners will often like to stay in more frequent contact with their wholesale wine suppliers, who can alert them to changes in product availability, new offerings or trends, and information they can use to grow their restaurant business. In your effort to develop and strengthen the relationship with your clients, be respectful of their time and have a reason for contacting them. We recommend that you use a spreadsheet like the one in figure 4.1 to keep track of your accounts.

SERVICE YOUR ACCOUNTS

There is an old rule in the sales business that 20 percent of distributors will account for 80 percent of your sales, and it is much easier to make sure those top customers are getting the service they need when you take advantage of modern sales technology and business analytics. These technologies can also help you maximize your relationships with your top wine club members, and there are even programs that allow you to use consumer insight data to evaluate your competition and find new opportunities for sales or distribution. Today consumers volunteer their opinions about products right on their mobile devices. In today's wine sales world, it is possible to take more than five hundred thousand points of data and use it to answer your most pressing marketing and sales questions about how and where to launch your products, and it can help you negotiate with distributors using consumer behavior data. Technology has advanced so that migration to newer technology tools is easier than you think. Following are examples of tasks that you may not initially consider in the management of sales relationships with distributors but that are easily tracked using technology:

- Managing trade spend for optimal return on investment
- Setting spending allocations according to approved budget plans, including depletion allowances and bill-backs
- Planning, budgeting, executing, and auditing all trade promotional spend down to the account level

ANALYZE THE RESULTS

Reliably predict which accounts are the best fit for your products by analyzing your sales. Using account purchasing patterns, technology can predict (and learn more over time) what products will sell in what accounts. This provides suppliers and distributors a powerful way to reach sales goals sooner with a higher certainty of reordering from you. In order to identify and target key accounts, with the understanding that all accounts are not created equal, you can use qualitative custom research that matches the best high-volume and high-image accounts with your SKU sales strategy. Reps can sharpen their focus on their most attractive accounts. Do your own research and make survey results actionable and automatic. To get the most from data collection, surveys must drive immediate action. Using tools that automate action means sales and marketing respond to changing market conditions at the right time.

VIEWPOINT

Given the importance of DTC sales in the wine industry, one might assume we have found the holy grail of customer relationship management. But compared to other hospitality industries, such as cruise lines and casino gaming, we fall far short of best practices. Sure, we have strong wine clubs with members that respond to our e-mail offers (at least some of them do), but there is so much more we could be doing to create ambassadors for our brands.

Let's begin with a few underlying assumptions. Do you think your top-tier wine club members are synonymous with your best customers? Are they the most satisfied with their tasting-room experience? Is their event experience exemplary enough that they bring more of their friends to each new event? Within your competitive set, is your wine club better than other clubs or lists to which your best clients belong? Are your top customers more satisfied with your product? Are they buying more frequently than before? Are they visiting more frequently than before, or has their profitability declined because they continue to taste and attend events without incremental purchases? Do you even know the answer to that last question?

We have the tools, and we have the data. The building blocks are well in place. But we fail to use the existing data to better understand, segment, and personalize our marketing programs. In the end, we could grow profits much further and create far more enthusiastic customers.

"BEST CUSTOMER" DEFINITION

Other industries, including the gaming industry, offer tiered club programs, but they are typically enhanced with revenue metrics to ensure that the best customer is indeed the most profitable to the company and, therefore, should receive the most club benefits. It's a sophisticated management system that allows companies to fine-tune their customer relationship management well beyond what most wineries attempt.

Wine club attrition is a significant issue in the industry, with the average wine club member leaving after eighteen to thirty months. Most wineries accept this as unavoidable or as merely the cost of doing business. But linking purchases over time to customer value can provide helpful insights into those members who increase or decrease their spending over time. By identifying the declining best customer and implementing personalized programs to avoid defection, you can make a measurable difference in your wine club profitability.

In the wine industry, our top customers typically "self-select" into our top-level program through their case commitment. In many cases, this is a good enough measure of who is our best customer, but let's take that a step further: a more robust measurement would include overall revenue, club referrals and affiliate purchases, e-mail responsiveness, and event participation. Do you measure and track all these activities? Because if you did, you'd have a 360-degree view of the most important customers in your DTC program.

In the gaming industry, many top customers are clearly identified by their overall interaction with the company, and the top-level customers are treated to complimentary gifts, suites, and gaming discounts on a regular basis. The gaming industry understands how valuable these top customers are, and they treat them accordingly. We need to learn from that example and improve our unique value proposition and treatment program for the most profitable and "growable" members.

MEMBER BENEFITS FOR GUESTS

Hospitality companies often offer voucher programs in which members can extend their offer benefits to others. This encourages customers to bring their friends along, an important element in marketing to the key demographic ages of forty-five to sixty-five. That group loves to travel together. When a member arrives with a group of friends, you can bet the casino is going to lay out the red carpet for them all, because they understand the enormous opportunity to build their customer base inside a highly desirable, preselected group of clients. We don't see wineries extending member benefits to guests. They seem to think you should join the club to receive those benefits, but that's a very short-sighted approach.

Casinos and hotels understand and market to both their most valuable and most growable purchasers. Successful hospitality companies understand and use complete profitability and purchase trends as the basis for deciding who should be given special offers. They want to know who spends the most now, but they also want to track who is spending more each time they visit.

For a major event in Las Vegas, be it a boxing match or a headline show, the best suites at any casino are going to be allocated to the most valuable, or the most growable, players and their friends. In the wine business, you might choose to limit access to the last ten cases of your best wine instead of sending a mass mailing to your entire list. Those last cases should first be offered as a special perk to those who are most likely to continue to grow their relationship with you. Those are not just "good customers"; they are the absolute cream of the crop. Shouldn't the customers who are most likely to continue to grow their sales with you have the first crack at the new vintage offered this year? If they consistently bring new members, new buyers, new "promoters," aren't they more valuable than other members? How do you recognize these trends and use them to create real loyalty and enthusiasm? The argument is always "They're here all the time," "They've been members for life," "They're local," and "They're friends." Instead of taking these customers for granted, it's important to maximize your relationship with them by offering them special deals and more attentive service.

How can you do this? By treating your top-level customers with a degree of personalized service that goes beyond the standard e-mails and newsletters and by creating private experiences and communications to your top wine club members based on their overall contribution rather than just their club level. Through their high-level purchases and continued support, these top clients have shown that they are more than just good customers. It should be our job to let them know that we recognize them for this and offer them a level of personalized service and offers that others cannot expect. The ancillary purchases of best customers are an excellent opportunity to promote new releases in a warm and friendly way rather than through the standard mass e-mail to all members. If some of your top customers bought two cases of the sauvignon blanc last summer, and it sold out, you may think you don't need to do anything more for these customers: "They're already sold." But that's the wrong attitude. At the highest level of revenue value, the organization can do so much more, and those actions will have a substantial, positive impact on profits.

IN CONCLUSION

Reporting and analytics tools have improved with technology. Modern technologies make it easier to move away from antiquated tools like physical file folders that require too much effort to gain meaningful insight. To be competitive you must evaluate what drives your sales, studying your leading indicators through technology and sales analytics, and if you want to be competitive, you may just want to do it on your mobile device. Looking at past purchasing behavior using basic pen and paper won't get you where you want to go.

It is easy to benefit from the expanding amount of useful data and technology to be a more efficient seller. Developing meaningful reports can mean organizing and managing many data sets. This includes consumer insight data, white-label key account lists, shipment data, and invoice-level retail account depletions. Suppliers and distributors of any size can benefit from technology and can now cost-effectively gain the same advantages previously only available to larger wineries.

DISCUSSION QUESTIONS

1. What is the logic behind sorting existing customers into ABCD accounts?
2. Should A accounts always receive more of your attention, or should you focus on B accounts given their potential for driving new business?

Chapter Five

Direct-to-Consumer Sales

SELLING WINE DIRECTLY TO CONSUMERS

The goal of any winery marketing and sales campaign is to get the consumers in the market to buy the wine you are selling. That's obvious. But what isn't apparent is why end consumers (e.g., the people who actually drink the wine) purchase wine in the first place and how wineries can help them make their decisions. If you understand this, you'll be able to help your retail and restaurant accounts sell more wine, and that's a key part of consultative selling.

But there is a bigger audience we'll discuss in this chapter. The hottest trend in winery sales is selling wine direct to consumers (DTC). Large wineries want to increase their DTC sales, and many smaller wineries sell their wines exclusively DTC, where the profit margin is much higher than selling the wines into the three-tiered system. Of course, you can't sell half a million cases DTC, but there are a number of wineries both in Europe and the United States that sell more than seventy-five thousand cases a year DTC at full retail price for each bottle. How do they do it? That gets back to the question we raised in the first paragraph about why consumers buy wine. In a recent study by Thach and Chang (2015), American consumers listed "taste" as their single most important reason for buying a particular wine.

This shouldn't be surprising since it's hard to imagine that consumers would want to buy a drink they didn't enjoy. But note that last category in figure 5.1: very few consumers buy wine to taste it critically. When you add all the reasons together, the summary seems pretty straightforward: consumers buy wine to have a good time with their family and friends. Despite what most people in the wine business seem to think, consumers don't buy wine to learn a lot of details about wine.

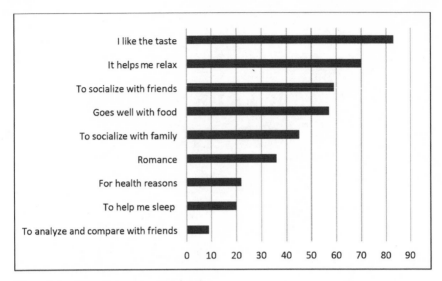

Figure 5.1. Why Consumers Drink Wine

And if consumers do buy wine because they like the taste, it would be a good idea to find out what kind of taste consumers want. Happily, Thach and Chang (2015) asked that question as well. And figure 5.2 shows us what consumers told them.

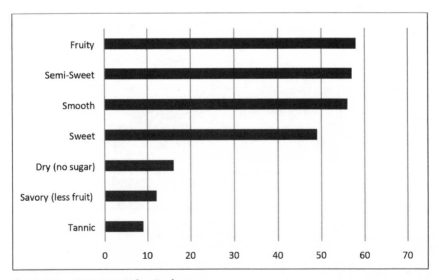

Figure 5.2. Consumer Wine Preferences

Of course, this sounds very much as if the survey focused on "beginning" consumers, not more serious wine drinkers. But here is how the respondents described themselves regarding their wine knowledge: "only 4 percent of the sample rated themselves as wine connoisseurs or experts, but 21 percent said they had advanced knowledge, meaning they considered themselves to know more about wine than most people. The majority ranked themselves as having intermediate wine knowledge, at 57 percent of respondents, and 18 percent identified themselves as wine novices."

Wine columnist Lettie Teague (2015) of the *Wall Street Journal* notes this trend: "What's the most common word oenophiles use—and misuse—to describe a wine? The answer might surprise you. According to several retailers, it's 'smooth.' We want to stress it is oenophiles—serious wine lovers, who are being quoted here, not greenhorns." If that is what consumers want, isn't that what wine salespeople should be talking about? Selling wine to consumers is just like selling wine to the trade. You do your research to find out what they like. You open the conversation by asking questions and listening to the answers. It's an outline very much like the one we will present for that opening sales call to the trade.

In this case, the scenario might be in a winery tasting room rather than in the office of a wine buyer, but there are definite similarities in the process. Below we outline how author Paul Wagner teaches this subject in his classes at Napa Valley College.

Greeting

This is your opportunity to make a good first impression, and you don't ever get a second chance. When customers walk in the door, they should be welcomed within seconds with a warm and professional greeting. Whether the tasting room is jam-packed with people or empty, the customers want to feel as if their visit is important to the winery. Your cheerful greeting will set the stage for the rest of your interaction with this customer. If customers have to wait forty-five seconds for anyone on the staff to say hello, they may well feel that they'll have to wait for everything else as well. And they won't feel that they are valued customers.

Probe for Their Needs

After the greeting, you should follow up with a series of two to four short, open-ended questions that will help you understand the mood, interests, and expertise of the customer. These questions lay the critical groundwork for two-way communication, because you're not just talking, you're asking questions, and you are going to listen carefully to the answers. Now you have a conversation, and you are not just making a sales pitch. Closed (yes or no) questions won't give you the same quantity or quality of information. These are questions you should ask:

- How are you today?
- Is this your first time at our winery?
- What other wineries have you visited today?
- What kinds of wines do you like?
- If the customer is wearing a T-shirt from a local university, or from a local wine festival, it makes sense to ask about that, too.

You are not just making conversation here. Each of these questions should give you added insight into who the customer is. If she tells you that she has visited many times before, that is key information that should set off a series of events for your "valued repeat customers." If he tells you that this is the first winery he has ever visited in his life, that's a completely different situation. If she likes "red wines," that may indicate a certain level of wine knowledge. If he mentions some of the lesser-known varietals of the Rhone, that shows a very high level of wine knowledge. And if you are unsure, you can always continue to ask open-ended questions along the same lines. By doing so, you demonstrate an interest in the customer. That sends a positive message on its own. And you also add to your file of information about the wine drinker standing in front of you.

Build Rapport

Asking questions is only half of this part of the sales call. Every time you get a new piece of information, you should respond by sharing something that indicates you have heard what the person said, that you share some of the same perceptions, and thus that you have something in common. Did he visit a local restaurant that recommended visiting your winery? Share your appreciation of the chef there, or explain your relationship to the waiter. Did she attend the wine festival? Tell a quick anecdote about your experiences there.

In every case, your goal is simple: "I heard what you said. I share some of the same feelings. We have this in common." These are the necessary steps to building a foundation of rapport with your potential customers.

The Pitch

Now that you know what kinds of wines the potential customer likes, and even possibly which brands, styles, or stories he or she prefers, you can present a few of your products as being worthy of attention. Always base this initial offer on what you have learned from the answers to your questions. "If you like red wines, maybe you'd like to try a taste of our merlot?" "Is this the first time you've visited our winery? How can we make this visit a good one for you?" "Would you like to go on a tour to see how we make the wine, or taste a few of our products?"

We like to think of this part of the sales call as walking arm in arm through your offerings. You now have an idea of what the person likes, and you are walking with him to show him the best candidates for his purchase. While you do this, it is critical to keep listening. Did you get the first part right? If you're talking to a group of customers, did you get answers from all of them? Does only one of them prefer red wines and the rest perhaps prefer whites? When she said "dry wines," did she mean not too sweet? By "walking" potential customers through your best suggestions for purchase, you can fine-tune your pitch to find the perfect match. And never stop asking questions. Every question leads to more information, and that information can help you with the next stage.

The Negotiation

This is where you begin to fine-tune the order. If someone mentions a wedding or a graduation, those are special occasions and might encourage you to pull out an older vintage or make a special offer on a large-volume purchase.

At the same time, you may find that your customer doesn't like red wines after all, at least not your red wines. Or he doesn't like the color of your label. It's time to address some of the potential adverse reactions. Some wines taste better with food. If you can offer a small taste of something that alleviates the tannins in your red wine, you may gain a customer for life. If you can create in your customer's mind a mental image of drinking the wine in an ideal setting, you might convince the customer to buy a bottle for that very occasion.

Some negative responses can be overcome with the right turn of phrase or logical argument. "Our wine isn't too expensive if you really appreciate the time and love that we put in each bottle. Save it for a special occasion." Sometimes it is better to cut your losses on that wine and either move to another option or admit that this customer doesn't like your wines. Suggest an alternative and spend your time with a customer with more sales potential. Working these issues out with a customer can be fun, even exhilarating, for a top salesperson.

The Close

Now that you have explored all the options with your customer, it's time to close the deal. Don't be afraid to be direct about this. By following the steps we've shown you, you have developed an excellent working relationship with the customer. Trust that relationship and ask for the sale. There are lots of ways to do this, but they all boil down to a few simple questions:

- So, how many bottles do you want to take home today?
- That's a bottle of the merlot and a bottle of the chardonnay. Anything else?
- Shall I add in a bottle of the reserve for the rehearsal dinner?
- If you buy two more bottles, you get a discount. That's like getting the second bottle almost free.

The Promise

The customer now has his wine, and you have his money. That's not the end of things. Why do people visit a winery rather than just buying wine at their local store? They enjoy the tasting and the conversation. They love hearing the stories. And they love meeting someone in the wine business they can call a friend. What you have done, by building that consultative relationship with your customer, is something that should not end when she walks out the door.

Think of this encounter as a first date. Give the customer a reason to call you again. Give him your business card, and invite him to call ahead next time so you can give him a special tour or open a special bottle for him. Ask for her contact information so that you can let her know when the new vintage of her favorite wine is released or when the winery is running out of the wine she loves. Make a reservation for him at a local restaurant. Make them part of the winery family, and they will continue to be essential customers.

The Follow-Up

This part is obvious. You need to deliver on the promises you've made. If you offered to tell him about a new release, make sure you do that. Ask the restaurant where she is eating lunch to give her special treatment as a friend of yours. Give him another reason to call you back. If she has bought more than six bottles of wine, give her a call in a couple of weeks and ask her if she is still enjoying the wine, making sure she got it home safely and has no concerns about her purchase.

Your goal is not just to sell customers wine; it is to make them part of the winery family. In the best of all possible outcomes, you would like customers to become more than just customers. You want them to tell all their friends about your wines. You want them to become missionaries for your wines!

A FEW WORDS ABOUT HOW *NOT* TO TALK TO CONSUMERS

Wine consumers are overwhelmed. We know this because we see it every day in our stores, our restaurants, and our wineries. Even a wine expert can find him- or herself at a complete loss in front of a restaurant wine list with hundreds of wines in each category. And the wine business makes it harder for consumers to fall in love with wine. In the advertising business, there is an old adage that says, "Talk about benefits, not features." In other words, don't talk about what the product is; talk about what it does for the consumer. In the world of wine, we do exactly the opposite. We talk about every possible feature of the wine, and we never talk about what the consumer wants: to have a wonderful time with family and friends.

We talk about points and ratings and medals as if they were priceless. But a quick survey shows that nearly seven thousand wines received a rating of more than ninety points from the *Wine Spectator* tasting panels last year. One hundred twenty-five wines from Napa alone got scores of ninety-five points or more. *Robert Parker's Wine Advocate* gave 160 wines from the 2012 vintage ninety-six points or higher. Each year the scores of wine competitions give out gold, silver, and bronze medals to thousands of wines.

Consumers' eyes glaze over at this kind of thing. They appreciate the fact that your wine is good—and your independent confirmation that it is good. That's not why the vast majority of consumers buy wine.

The market today is saturated. Your competition is overwhelming. Consumers are overwhelmed. And in response, the wine industry seems to think that consumers want to know every single detail of how a wine is made. In short, in addition to too many wines and too many scores, we believe consumers want more information, too. They don't. They want a reason to fall in love with a wine.

People don't fall in love with facts. They fall in love with stories. An old saying in the wine business is that "facts tell and stories sell." If you want customers to remember your wine, you had better stop talking about water retention and calcareous soils and start telling them something that touches their heart. You can't intellectually convince people to like a wine, but you can tell them such an endearing story that they want to drink it. In short, the best way to talk about wine is to tell a good story.

Here are some guidelines on talking to consumers: don't provide tasting notes for your wines. Cornell University (Thomas et al., 2014) did an excellent study that shows that most consumers don't like the tasting notes. Consumers read the long, complicated descriptions and then admit that they don't find the same characteristics in the wine. They then conclude that the wine is not for them. In a tightly controlled study, price sheets *without* tasting notes outsold price sheets *with* tasting notes by 40 percent in both New York and Napa Valley tasting rooms. And yet wineries still insist on providing tasting notes for their wines.

Don't talk about AOCs, DOCGs, DOCas, limited yields, viticultural practices, or malolactic fermentation. Consumers really don't care about these details. They are too complicated, too hard to remember, and don't offer the consumer anything of value. Do you know what kind of guitar strings Eric Clapton uses? You don't. But you love his music. That's a good example for the wine industry. Stop talking about how it is made, and talk about why you make it.

Don't teach chemistry. Consumers buy wine so that they can take a short vacation via the bottle of wine. They do not want a vacation to their high school chemistry class. The romance and charm of wine have nothing to do with pH or titratable acidity.

Don't teach geology. Consumers don't need geology lessons. The advantages of calcareous or volcanic soils do not move them to rapture. No consumer has ever walked into a wine shop and asked for a bottle of wine from volcanic soils.

Don't teach botany. The difference between rootstock 1103 Paulsen and 5C adds little to their dreams of the wine country.

Whatever you do, don't teach enology. The differences between Slovakian, Slovenian, and Slavonian oak should remain a mystery. Wine is the most romantic beverage in the world. Every consumer knows this. If you think that cold soaks, indigenous yeast, and malolactic fermentation are romantic, you are mistaken.

Wine buyers want a story. They want to fall in love. They want to open a bottle and take a journey—a journey to a beautiful and interesting place, where people live not for mortgage payments and school report cards but for the quality of life itself. Tell them that story about your wine. Tell the story of the people who make it with love and attention.

What makes a good story? For a start, stories are always about characters, not about things. As humans, we need stories to understand the world. Most wineries would be much better served explaining *why* they make the wine they do rather than providing endless technical explanations of *how* they make that wine.

Does the story you tell about your winery or wine meet this standard? If not, get to work and refine it until it does! Use that story to create an emotion around the brand: talk to their hearts, not to their heads. Here's a good example of what we do wrong in the world of wine:

Would you like to know more about her? If she were a wine, this is how we would describe her.

She is made of the following elements:

- Oxygen: 61 percent
- Carbon: 23 percent
- Hydrogen: 10 percent
- Nitrogen: 2.6 percent
- Calcium: 1.4 percent
- Phosphorus: 1.1 percent
- Trace elements: 0.9 percent

And yet that is not what we want to know. We want to know who she *is*.* We want to know what she does in her spare time. We want to know what makes her laugh and if she likes wine. If you want to sell wine to consumers, you need to stop spouting statistics and schematics and start telling stories that capture their imagination.

Sadly, many wineries are not run by people who understand the sales process. In most cases, they are run by people who make the wine. Making wine is not the biggest challenge in the wine business—selling wine is. And most wineries struggle to do this effectively. We take a lighthearted approach to this topic below.

MISTAKES WINERIES MAKE

We often lack a clear and memorable key message, and if we do have one, we don't communicate it. In fact, we pretty much say what every other winery says about itself. And then we complain that consumers can't tell us apart.

We make it hard to visit and provide little help via our website, telephone, or app. Hospitality is based on making these things easy for your guests. If your website isn't mobile friendly or doesn't have basic information for visitors right up front, you can't expect them to visit you.

The first impression is our parking lot. That's not good. We spend a lot of money on our label design . . . and then have a gravel parking lot with muddy puddles in the winter. We don't make people feel comfortable; we make them feel ignorant. No other industry spends as much time "educating" their consumers, which is why other industries are often more successful. Coca-Cola manages to sell six trillion servings a year without telling people how the product is made.

Our staff is very knowledgeable about wine, and they act like it. Remember that Emile Peynaud, the great Bordelais wine expert, clearly stated that he could not distinguish the wines of the various communes of Bordeaux in a blind tasting. Let's stop talking about this kind of thing. And let's stop making our customers feel stupid because they can't either.

We teach instead of talk to our customers. The music industry doesn't do this. Why do we? You don't have to know how to make a piano in order to

---

* The photo is of Meritxell Falgueras, who writes about wine in a wide range of publications, such as *Time Out Barcelona*, the magazine of *La Vanguardia*, and *Vinos y Restaurantes*. In 2007, she won the coveted prize Nariz de Oro Joven Promesa de Cataluña. Since 2007, she has hosted a weekly show about wine, *El día a la COM* on COM Radio. In 2010, she debuted as host of *Vins a Vins* for BTV in Catalonia and published her first book, *Presume de vinos en 7 días* (Ed. Columna/Salsa Books). She was named the Queen of Verdejo in 2008, her "Wines and the City" blog won the prize as the best wine blog of Catalonia, and *Esquire* named her the top sommelier in 2010. She is currently producing the second series of *Vins a Vins*. Meritxell has changed the way Spain talks about wine.

enjoy Chopin or Elton John. And you don't have to understand malolactic fermentation to enjoy a bottle of wine.

We always give the same tour, no matter who is in the group. It doesn't matter if the customer is a master sommelier, a master carpenter, or a millennial coder, which means that the tour is never right for that customer.

We provide tasting notes despite definitive studies that show that tasting notes in winery tasting rooms reduce wine sales significantly. We can't stop ourselves. When we tell consumers that a wine tastes like cassis and *garrigue*, we are not doing anyone a favor. They assume that they are not smart enough to understand our wine and leave without buying a bottle. But we feel brilliant!

SALES TEAM COMPENSATION

Over the years, Paul Wagner has consulted with many wineries concerning the compensation package they offer to their tasting-room salespeople. The goal is to compensate your tasting-room staff so that they are encouraged to sell well and foster long-term, high-volume relationships with your visitors. Compensation should be based on a clear understanding of the operating costs of your tasting room and should only reward employees after sales have reached the point where the winery is making a profit.

Begin by analyzing your costs of operation for the tasting room. This should include all costs and expenses, from rent and insurance to wages and benefits. In the most straightforward format, you can divide this annual number by 365 to get a daily cost of operation for the tasting room. This is your base number—probably something between two hundred and two thousand dollars a day, depending on your facility and staff costs.

For a more sophisticated analysis, you can fine-tune this number so that it is higher during the summer, lower during the winter—or even higher on the weekends and lower during the week—to reflect your varying staffing costs.

This base number is the foundation of the commission structure. It is the figure you must reach each day for the tasting room to show an operating profit.

Now calculate the sales volume you must sell to make a profit on any given day. To do this, you should compare your cost of goods with the retail price of the wine. This will provide you with a specific percentage of cost of goods and profit margin for each wine, but what you want is an average—a general percentage figure that you can use for all wines. This cost of goods usually ranges from about 35 percent to 50 percent.

Ready to make the first calculation? Let's use an example. If your operating costs are three hundred dollars a day and your cost of goods is 50 percent (which is pretty high), you need to sell six hundred dollars' worth of wine per day to make a profit for the winery. That figure becomes your sales goal. Once you reach it, you are making money. For example:

Total sales in any given day:	$600	$720
Minus operating costs per day:	$300	$300
Minus cost of goods (50 percent of sales)	$300	$360
Net profit/loss:	$0	$120
Commission:	$0	$12

When do you pay commissions? You pay a 10 percent commission any day that the sales team in the tasting room goes over your sales goal. However, you *only* pay the commission on the volume above your sales goal. So, if the sales one day are $720, you pay a commission on the $120 over the sales goal. Yes, it is only $12 (10 percent of $120). Only those employees who work that day get that commission.

There is not much money in this example. However, it adds up. Moreover, it works. Your staff will be more willing to open the doors a few minutes early to start the day with a big sale and get up to that sales goal immediately. They will also be more willing to stay late because they know that every bottle they sell gets them a commission.

Moreover, they will be more willing to make a phone call to a good customer because that telephone sale gets added to the commissions of that day. You'll even find that they spend their free time making phone calls to customers, visiting other wineries to encourage referral business, and doing all they can to increase their sales volume. This compensation structure also encourages staff to work on the weekends, when the sales volume is highest, or to work with the best salesperson—they want his or her help on the commissions!

Yes, there are flaws in this system. Not every staff member sells the same amount of wine during the day. However, if one is washing the glasses so that the other can sell more wine, both employees should share in that commission. This builds teamwork.

Yes, there will be days when you pay a commission after three straight days of not reaching your sales goal, meaning that in the big picture, you will be losing a little more money. And yes, your sales staff will probably try to manage their telephone sales so that they all happen on a day when the sales goal has been reached.

But those are short-term problems. Your staff is going to focus on making sure they reach the sales goal every day. At that point, these concerns are irrelevant. The very best part of this system is that you only pay commissions when you are already making money.

How big can this get? It can get very big. I know one winery owner who once met with the best salesperson in the tasting room to tell him that the commission structure was too generous. Why? "You are making more money than I am," said the winery owner. "I sell more wine than you do," replied the tasting room staffer.

The commission structure stayed.

Think about it. If your staff surpasses the sales goal by a total of one million dollars, they will take home one hundred thousand dollars in commissions. That's a *lot* of money. So is the roughly five hundred thousand dollars of profits your tasting room will be making at that level.

Section II

THE CONSULTATIVE SALES PROCESS

The chapters in this section are specifically focused on selling to organizational buyers. Though all wine sales should be consultative in nature, selling to buyers who will, in turn, sell the same wine to their consumers is more complex due to the volumes purchased and the risk that the wines purchased won't resell quickly at the price needed.

These chapters will place under the microscope a recommended sales process gleaned from countless interviews with wine sales professionals and the customers they wish to attract. Often we will include comments gleaned in these interviews as a means to illustrate the steps in the process. Deconstructing the consultative selling process across multiple chapters, with accompanying exercises, is intended to give you—the reader and future wine sales manager—a clear understanding of best practices and the ability to execute on each across the full spectrum of the wine trade.

But let us make one thing clear: selling to the trade is the same basic process as selling to consumers. And while the trade has an additional level of complexity, understanding this process for the trade will also make you a better salesperson to the consumer. Knowing the market, understanding customer needs, overcoming objections, closing effectively, and excelling in sales follow-up will increase your sales in every scenario, from the winery tasting room to the large corporate account.

Chapter Six

Buyer Motivations
and Presales Call Planning

THE NEED FOR A FLEXIBLE
SELLING APPROACH FOR THE TRADE

Like you, your customers want to be successful. Your success as a salesperson depends in large part on your commitment to your customers' success and your ability to help customers make decisions that contribute to their success. As wine distribution sales professionals, this means helping your customers create the right mix of wines to have available that will attract customers and be quick to resell at a profit.

In a winery tasting room, this can mean making sure the customers are happy and comfortable in the future enjoyment of the wine as they serve it to their guests. In the trade environment, it's more complicated than that. A lot can be riding on the purchase decision your customer makes for his or her organization. Wine specialty retailers operate on razor-thin profit margins, once you take into account their costs per sale. If they stock their inventory with slow-to-move labels and vintages too often, their future as a viable business will not be particularly bright. The same is even more important for distributors and other types of wine retailers. Also, adding new wines must generate new sales, not just cannibalize what is already in the inventory. Hence, the viability of their business is dependent on making wise purchase decisions.

Organizational buyers will attempt to reduce or minimize the risks in one of two ways: (1) they will repurchase from a supplier the wines they know and trust or (2) they will shop around for competitive options. Not unlike retail shoppers in a wine shop or restaurant, organizational buyers like to find suppliers they can trust, and they depend on building, in a sense, customer loyalty. However, companies—and sales managers—should never take their customers for granted. Just because your firm has been servicing a restaurant

67

group's account for years does not mean that you are without competition. Buyer-supplier relationships are economic exchanges, so buyers can and do switch suppliers if they believe it is in their organization's best interests to do so. Distributors, as resellers, naturally purchase wines from many producers. Moreover, most restaurant groups that sell large volumes of wine will work with multiple distributors, not only to source new labels but also to ensure that they receive the best value for their purchasing dollars.

Some brands and some wines practically sell themselves, regardless of the competition. They have such good quality, successful marketing, and effective prices that vintages are claimed before they are ready to be shipped. That is what every buyer wants. It is what distributors want, and it is what retailers want.

Such wines are few in number and have limited supply, and this book is about how to sell the rest of the 130,000 wines in the market. Part of that is convincing buyers that some of the elements are in place that will work for them. Some of it is getting the rest of the elements in place. Some of it is convincing the buyer that there is a unique role for this wine to play in the larger scheme of things, via pricing, varietal region, style, or the wine's sales performance in other markets. This requires gaining an in-depth understanding of your customers' needs and the circumstances behind their needs.

The challenge in selling wine to a retailer is that you have to fight for shelf space because any new product is going to displace an existing one. There is

Figure 6.1. Be Committed to Your Customer's Success. "We need to help the customer make good business."

a finite amount of shelf space in the store, and getting it is often considered a zero-sum game among buyers. Selling your product has to be more profitable than selling someone else's product. Your wine has to increase sales and/or be sold at higher margins to be profitable. To become trusted sellers to retailers (or distributors), wine salespeople face the same kinds of challenges: a limited or finite amount of time on the part of prospective buyers. And so you have to convince them that spending time with you is more helpful and profitable than spending time with your competition.

THE FUNDAMENTALS

Customers will tell you how you can help them succeed when they express needs—that is, when they express a desire to improve or accomplish something. A need is a *desire to accomplish or improve something*. You help your customers, your organization, and yourself succeed by understanding and satisfying your customers' needs.

IDENTIFYING CUSTOMER NEEDS

You cannot see a customer's needs. They're "inside" the customer, and only the customer can tell you what they are. That's why we suggested the earlier series of questions for retail salespeople as a way of not only breaking the ice but also learning about the customer. And in distributor sales, there is even more to learn. For example, you can be reasonably assured a customer has a need when he or she uses the "language of need"—words and phrases that express desire:

- I want . . .
- Our strategy is . . .
- What we need is . . .
- It is important to us . . .
- I'd like to see . . .
- I hope to . . .
- What we are looking for is . . .
- We must find . . .
- We are interested in . . .
- What matters most . . .
- I need to find . . .
- My objective is . . .

For consumers, some of these same issues arise and are identified in the Constellation Genome project, as mentioned in chapter 3.

When you are on a sales call, it's important to listen for and recognize the language of needs. If you don't, you might make unwarranted assumptions about what a customer is looking for and waste time talking about things he or she isn't interested in. As previously mentioned, it should come as no surprise that all distributors, retailers, and restaurants have, first and foremost, two overarching needs: the need to purchase only those wines that (1) will bring customers to their stores and (2) are quick to resell to their own customers. Just like any other industry, your organizational buyers have little incentive to substitute one wine for another if there is not a net gain in sales. In addition, your buyers do not want to get stuck with inventory that does not move. A key element of salespeople's job is to provide convincing evidence that the wines they can provide will increase sales and at a good profit margin. If you cannot convince them that you can alleviate these principal concerns, nothing else you can say or do will persuade them to buy from you.

However, there is nothing we can do or say that can *guarantee* a case of wine bought today will be resold at a profit tomorrow. Savvy distributors, retailers, and restaurant managers understand this and will look for wines and supportive services that can fulfill needs that, when combined with their own expertise, will contribute to their success in selling to their customers. Successful salespeople will put in the time to have evidence at the ready to support their customers' needs. The following are needs often expressed by retail customers:

- Price: A restaurant selling wine by the glass will often try to recoup its cost with the first glass sold. Though quality is important, the restaurant owner or manager will generally be reluctant to purchase a wine for this purpose that extends beyond a certain price point. Grocery stores are also narrowly focused on price, knowing that price sells. In Europe, the sales volume of €4.99 table wines is twice as high as for €5.99 wines.
- Wine ratings: These are scores assigned to a wine by one or more wine critics who are opinion leaders in the wine industry. These are the critics' subjective evaluation of the quality of a wine and are generally provided using a numeric score (e.g., a twenty- to one hundred–point scale). A restaurant manager may express to a wine sales rep an interest in considering wines that score above ninety points that are priced below $120 a case. Thus the rating, in combination with the price, gives assurance to the customer that the wine will be perceived as a good value by their downstream customers.
- Purchase incentives: These often take the form of financial incentives to customers who are willing to enter into an exclusive buyer-supplier rela-

tionship with a producer or distributor for a fixed period of time. We have witnessed such incentives taking the form of a wine distributor purchasing wine glassware for a new restaurant client in return for an exclusive contract with minimum purchases. Another example is that grocery stores prefer distributors who provide merchandising services that place the burden on the vendor to keep their shelves stocked and in good order. These incentives save money for their respective customers, who may then choose to pass along some of these savings to their downstream customers to remain price competitive among their competitive set.

- Exclusivity: Most buyers want to carry in inventory what no one else in their market offers. A restaurant will understandably be reluctant to carry a wine label that its customers can readily purchase from a grocery store or big-box wine store for 70 percent less. Moreover, a restaurant may be willing to experiment with a new wine if there can be a guarantee that if it is later determined to be a big seller, the buyer can be assured that it will not be sold to their competitors. Though making exclusive offers is considered to be anticompetitive in many US states, placing a sales order on a limited inventory of labels that are distributed over a fixed period of time to a single buyer is a common legal practice entered into in the give-and-take of price negotiations.

- Customer support: Many restaurants have small back-of-the-house storage facilities that limit their capacity to store a wide selection of wine. As such, they will often express a need to purchase less than a full case of certain labels and a need for frequent deliveries and last-minute help when certain wines run out. Such support services—though costly—ensure that your customers can provide a relatively uninterrupted supply of wine to sell regardless of unanticipated spikes in demand. In addition, grocery stores and big-box wine retailers often request merchandisers and promotional support (which will be discussed in detail in chapter 14).

When all things are equal among competitors regarding wine quality, price, and service, probing for the circumstances behind your customer's needs may tip the balance for you in winning the business. These queries are not designed to be manipulative since buyers will quickly see through such tactics. Instead, they are designed to be consultative, as you and your customer are working in a mutually beneficial way together.

Consider for a moment the following consultative approaches some of you may already use in some form. The framework we present them in will hopefully spur creativity to consider those you do not use and, better yet, develop new consultative strategies of your own.

- Thought leadership: this can take many forms in wine distribution sales. We have witnessed former chefs who run successful wine distributorships that work with their restaurant clients on food-wine pairings. Fine-dining restaurants survive and thrive on both food and wine sales, and finding ways to expand the synergies between the two creates win-win opportunities for both the buyer and seller. In addition, wine reps who commonly conduct a gap analysis (see chapter 9) for their clients, examining their wine inventories for missing varietals, styles, regions, and price points that sell well in other markets, are examples of sales reps as thought leaders. By focusing on his or her customers' underlying needs to change or constantly improve their portfolio to keep up with consumer trends and improve sales, the salesperson is differentiating his or her products as something of unique value and taking a dual focus off price and exclusivity.
- Cultivate an emotional or personal connection with a wine that builds enthusiasm for the label. Providing a client a tour of a vineyard where he or she meets the owners and can experience firsthand the pride and tradition in its production not only deepens the understanding the buyer has of the wine but can also create a sense of brand evangelism in the buyer as he or she resells the wine to customers. In addition, wine reps who are willing to share such experiences in training a restaurant's waitstaff on a wine and its food pairings are putting into practice storytelling and the emotional appeal it can evoke. We have also found that sales reps tend to sell more wine when they have a deeper connection, in part because the personal story they can tell about the wine and winery imparts a degree of enthusiasm and emotional appeal to their clients and downstream customers themselves. Hence growers and importers can serve an important role in educating their distribution partners' sales reps.
- Challenge your customers to rethink their customers' status quo and teach them something new and valuable. When it comes to wine, there are no real innovations. Instead, there are trends in consumption that vary from country to country and region to region. Prosecco may be selling well in the southeast region of the United States, while pinot grigio is popular in the United Kingdom. Like all trends, wine tastes are constantly changing, and the rate of change is speeding up. You may not always find customers who express a need or wish to change their status quo. However, the wise sales rep will come prepared to challenge buyers' status quo when necessary—when they may be overlooking opportunities their business can exploit. We will examine this in greater detail in chapter 10 when we discuss supporting your customers, but just as a sales rep is a source of marketing intelligence for his or her company, he or she can also be a source of marketing intelligence for clients. Too often one's clients are so

overly embroiled in their day-to-day operations that they are not aware of their competitive position or of recent trends in wine offerings and sales. Wine reps who have a sincere interest in helping their clients succeed have an obligation to teach and occasionally challenge their clients to help them do better.

WHAT IS A SUCCESSFUL SALES CALL?

A successful sales call is one in which you and the customer make an informed, mutually beneficial decision. The sales call itself may have been initiated by the customer (e.g., inbound sales in a tasting room, restaurant, or retail shop, or a phone call from a trade customer) or by you, the salesperson (e.g., prospecting, account management). As a wine sales professional, it's your job to direct and manage the call so that this outcome is achieved.

You reach mutually beneficial decisions through an open exchange of information that focuses first and foremost on the customer's needs. Learning the skills of consultative selling will help you facilitate such an exchange on a call.

ELEMENTS AND FLOW OF A CONSULTATIVE SALES CALL

There are many models that highlight various recommended elements of a sales call. The consultative selling method is but one of them. All have their similarities but also differences. However, all have a beginning, middle, and end focused on establishing a positive impression with the customer, discovering what the customer needs or is attempting to accomplish, presenting one's product and support service relevant to these needs in as favorable a way as possible, responding appropriately to any customer concerns, and asking for the sale. What consultative selling adds is an emphasis on benevolence, credibility, and ethics that, if sincere, increases the trust and respect you wish to develop with the customer.

Elements of the consultative sales process that this book advocates are described in the chapters that follow:

- Precall research (chapter 7): Understand the mood and expertise of the customer as best you can and his or her likely needs and interests in your product, and ready your proof devices.
- Call opening (chapter 8): Make a positive first impression, propose an agenda that includes requesting permission to perform a gap analysis of

the customer's portfolio, state the value of the agenda and subsequent gap analysis from the perspective of the customer, and check for acceptance.
- Probing the customer's needs (chapter 9): Confirm the customer's needs through good questions and listening, and conduct the gap analysis looking for missing varietals, styles, countries, and price points in the customer's portfolio that you can fill. For off-premise buyers, also consider missing elements in their merchandising and in-store promotions.
- Supporting your customers (chapter 10): Present relevant features of your product and services as benefits to the customer.
- Closing the deal (chapter 11): Advance the sale, ask for the business.

With *precall research*, you are preparing yourself for a two-way conversation with the customer focused on what he or she will likely want and what you can provide. To make your initial call with your customer both efficient and effective, much can be discovered through precall homework to determine such elements as the support services he or she will likely value, gaps in his or her inventory, and the level of his or her expertise in wine. These needs and preferences can later be confirmed or rejected during the probing stage.

The *call opening* begins when you first meet with your prospect and is composed of several stages:

- By establishing a positive first impression, you are communicating in appearance, manner, and deeds that you are credible, capable, and trustworthy.
- By proposing an agenda, you are communicating what you would like to accomplish during the call.
- By stating the value of the agenda, you are explaining what is in it for your client in allowing you to analyze his or her current inventory and guide the conversation as proposed.
- By checking for acceptance, you are asking for the customer's approval to move forward.

With *probing*, you gather more information about a customer's specific needs and preferences that could be gleaned in your precall homework. Often this means helping him or her find wines absent from his or her inventory that are selling well in similar locations.

With *supporting*, you provide information about how you can satisfy a customer's needs by proposing additions or changes to his or her wine portfolio and merchandising that will increase profitability. What wines are in your portfolio that can add value to the buyer's portfolio consistent with his or her wine strategy and other specific needs evoked by the customer during the conversation?

With the *closing*, you are asking for the consumer's commitment to the next logical step.

With *negotiating concerns*, you are responding to any concerns the customer has as they arise.

With *building the relationship*, you are attempting to strengthen the customer's satisfaction after the initial purchase. Ultimately, it is in the best interest of the buyer to reduce the number of wine distributors he or she works with, but only if his or her needs and interests are being met.

Selling to a lack of interest is essentially a plan B approach when you discover that a customer has no desire to change the status quo. Approaching respectfully, you probe for opportunities to evoke in a customer undiscovered needs that you can support.

Since customers usually consider more than one need when making decisions, the skills of probing and supporting may be repeatedly used during a call. Once you've opened the call, gathered information about a customer's need, and provided information about how you can satisfy that need, you go on to gather and provide information about additional customer needs. Then, at the appropriate time, you close the call. In responding to a formal request for proposal, the above objectives are addressed over multiple meetings, where the first meeting is used to identify the customer's needs, the second to present a proposal detailing how you can satisfy those needs, and the third to give a demonstration (e.g., site inspection) leading up to a closing.

IN CONCLUSION

Though wine is the core product that we sell, the product is complex in terms of varietals, styles, countries of origin, and price points. Add to this the merchandising, promotional, and logistic requirements that collectively make up a number of potential features that can be brought to bear in satisfying a customer's need(s). Successful wine salespeople first attempt to understand their customers' needs through their gap analysis and asking questions of their customers, then they present aspects of their products and services that can satisfy those needs. The old adage of consultative selling, "First understand before attempting to be understood," is one that all sales professionals should keep in mind.

Organizational buyers of wines have one basic need: to buy wines that they can resell to their customers quickly and profitably. However, there is complexity in their markets and business strategies that must be understood before you can effectively sell to them.

EXERCISES

1. Consider for a moment your favorite wine. Write a five-hundred-word story describing the wine and winery that you believe will be appealing to your buyers and their downstream customers.
2. Describe, or better yet conduct, a gap analysis of a local wine retailer.

DISCUSSION QUESTIONS

1. What is the outcome of a successful sales call?
2. How do you help yourself and your employer by helping your customer succeed?
3. A need is a _____. (Fill in the blank.)

Chapter Seven

Precall Research

Once you have a good basic knowledge of the depth and range of your wines and support services and how they fit into the context and competitive set of the current wine market, it's time to get started on making some sales calls. But before you jump into selling, we insist that you do your research first. You need to know something about each customer you are going to call so that you can provide him or her with the best and most accurate sales pitch possible. This is the purpose of those initial questions we proposed in our chapter on direct-to-consumer sales. But in the trade, there are more tools and more opportunities to learn about your customer before you make the call. It is absolutely critical to familiarize yourself with your customer, as essential as understanding your own products' place in the market.

A buying manager of a large wine retailer recently informed us that his secretary screens fifty or more meeting requests per week from wine suppliers. The few who are granted a meeting, let alone receive an order, are not the ones who lead in with a vague understanding of the buyer's needs. The ones who put in the time to research the potential customer's needs and wants, and have a clear value proposition that appeals to the customer, will get their foot in the door.

The same is true for wine distributors. Attempts to engage a potential customer by leading into the conversation with overly broad questions or vague value propositions will generally get you nowhere. Avoid openings such as these:

- What's important to you in taking on a new line of wines?
- We have a new label of chardonnay I believe you will like. I'd like to come over and leave some sample bottles with you to try.

As discussed in chapter 6, wine buyers for grocery store chains will typically base their purchase decisions on price, wine varietals and styles, growing regions, and delivery and merchandising services. In addition, wine distributors hoping to expand their wine portfolios are not only looking for new wines and price points but also for evidence to gauge a host of additional factors, including the producer's quality, reliability, and so on. Hence a sales rep calling on either one of these potential buyers should come prepared to offer products and services—supported by relevant evidence—focused on each of these basic needs.

With this said, the sales manager who puts in the additional effort to more thoroughly research the account can set him- or herself apart from the competition. Discovering that you are calling on a master sommelier, you may wish to bone up on the intricacies of your proposed wine and its history; better yet, bring a bottle for the customer to sample. In contrast, a prospective customer who has limited knowledge of wine may be better served by offering third-party proof of the wine's quality and popularity, such as awards, ratings, or sales in other markets.

The point here is that although there are common issues that each customer segment faces in making their wine purchase decisions, each potential buyer will no doubt have unique interests and preferences that it would be wise for the sales manager to address. Though not all these unique qualities can be discovered prior to meeting with your customers, many can be, and a knowledge of them can contribute to a more efficient and productive meeting when it takes place.

Consider for a moment the sales manager who has lunch at a bistro he or she would later like to call on as a prospective customer. Examining the menu and wine list and introducing oneself to the manager on duty, as well as viewing the layout of the restaurant, may provide unique insights about the bistro's strengths and weaknesses and how you may be of help to the manager as a wine supplier.

As a further illustration, consider a sales manager who spends time checking out the breadth of a wine store owner's inventory on the store's website or in a walk-through of the store. A wine store that specializes in luxury wines produced in its region will find little interest in your portfolio that falls outside this strategy.

In Richard Shell and Mario Moussa's book *The Art of Woo*, the authors offer several insights that can be adapted to wine distribution sales:

- Research your customer's situation: Analyze his or her goals, competitive position, and challenges, including what you know and what you need to find out.

- Confront the four barriers to your potential customer opening an account with you: getting that first meeting, establishing credibility, avoiding communication mismatches, and aligning his or her interests and needs with your products and services.
- Draft your pitch for requesting the first meeting: Knowing that you as a wine sales rep would like him or her to add your wine labels while dropping others, your prospective customer will need a solid reason to take your call.

In the best of all possible worlds, you should know as much about your customer's business as he or she does—because you want to be in a position to help him or her be as successful as possible.

Remember that every wine buyer in the market buys wine for only one reason: to resell it to his or her customers quickly and profitably. This is called *derived demand*. Your ability to help your customers meet the needs of their customers more effectively than other wine salespeople will not only help you sell more wine but will also create the kind of relationship that pays long-term dividends for both your company and your career.

This first part of your research is obvious: you need to know how much of your product your customer is selling. This should come from your own records and orders. By studying these, you can clearly draw conclusions about any buying patterns or practices that the customer has. Ask your company's other sales managers if anyone has had any previous experience with the account or buyer, and look for insights as to what value proposition would be best. Does she buy only one of your wines and not the others? Does she only buy your wines when she is offered special pricing? Does she tend to buy lots of wine just before the summer tourist season? And does she like to use products with strong merchandising support, or does she prefer to avoid those?

If your company doesn't provide one, you should develop a customer file for each account. In addition to all the obvious information like name, address, key contact, contact information, and a record of the sales (by week, month, and wine type) you have collected over the past few years, it should also include any notes that make the sales call easier. What day does the account prefer to see salespeople? What are the store hours? Does the buyer prefer you to make an appointment or suggest you drop in? An appointment is always preferred, simply because it's more professional. And even if the buyer doesn't ask for an appointment, you can suggest one as a way of being more respectful of his or her time. We also like the idea of noting any personal information that might be important. If the buyer is a huge sports fan, that offers an easy opening for your conversation and allows you to build rapport. But don't overdo this, as it can lead you to be less efficient and professional.

Once you have a sense of your own wines' track record with the customer, it's time to do a little more research. A visit to the store or restaurant will tell you what other wines the customer carries, including which of your competitors are in the shop. You can see how important pricing is. Does every wine have a discount note on it? What kinds of wines win one of the coveted floor stacks or end-aisle displays? Are wines with big point scores or gold medals given special placement? What kinds of food-and-wine pairings does the chef like, and which wines are featured in those? And make sure you notice where any of your wines are merchandised and how they seem to be selling. Are they positioned correctly? Will your client's customers be able to find them easily? And are they placed effectively in relationship to your primary competitors?

If you do this visit correctly, you should have a very good idea of what this customer wants regarding wines, wine styles, pricing, and inventory levels. That's a great basis to start developing a pitch on how some of your wines fit into your customer's business model.

But before you do that, it's time for another few visits—to your customer's local competitors. Stop in and see what is working for them. Get a sense of what wines and wine styles seem to be popular in the competition's stores and restaurants. If you can find something that is working well for the business down the street, you have a topic of conversation that can lead to a role as a consultative salesperson.

There is one more step you'll need to take before you are ready to develop the sales pitch. You need to study the data and market trends that show where the world of wine is going. If you can identify a few places where your customers may be missing an important and growing category, you will be able to both sell those products to the customer and enhance his or her long-term success: a win-win if there ever was one.

When sales managers share information about what labels are selling in other markets, what trends are predicted for the future, or how some wines may cross-merchandise to consumers of other wines, they are providing valuable context and assistance to the buyers—information that can make the buyers more successful. That is the essence of consultative sales.

Because on-premise accounts tend to be more specialized, they are generally more complex as a whole. And that means that sales techniques will have to be more carefully tailored to the specific needs and styles of the buyers. Some may be receptive to a sales rep who consultatively asks to analyze their wine inventory and run a gap analysis as to what may be missing. Identifying gaps in price points, varietals and styles, and wine-producing regions can help the restaurant increase sales and profitability by refining or optimizing its selections. Others may see value in a sales rep working with his or her culinary teams in achieving better food-wine pairings or in organizing training for their waitstaff on how to more effectively describe their offerings. Still others may be focused on the cost of pour (price points) at certain levels of wine ratings. Because they offer a more limited selection of wines, on-premise accounts are always more sensitive to responsive delivery service and reorder issues. When they run out of a wine, they want it restocked immediately, not next week after their big weekend.

Dine at the restaurant to study the wine list firsthand for gaps and better food and wine pairings. Just as you did for the retail shop, review the account file or CRM system carefully and/or probe your sales director to have a reasonably complete understanding of your customer's needs before contact. If the prospect is new, learn from studying the wine list and the style of the restaurant so that you can begin to understand his or her motivations and strategies. Once you've done that, you can begin to suggest how some of your products might fit into his or her plans. If there is a dish on the menu that you think is ideally suited to one of your wines, that's an excellent place to start. If you know about a trend in the market that is taking off, explain how that might work with one of your wines and offer to help your customer get started on a program that would take advantage of the trend.

Remember, even though you are tasked with increasing your sales to a list of existing and new accounts within your territory, your customers are busy and are buying with greater care and reluctance than ever before. There are more than 130,000 wines in the US market alone and salespeople to sell each one of those wines as well. That is dramatically changing the wine sales playbook. Navigating these different interests requires an understanding of a customer's *needs* before your sales call and how your product and service can produce *value* for him or her. Failure to generate value risks creating the perception that interactions with salespeople are a waste of time.

The point here is that retail wine buyers and restaurant beverage managers are busy individuals who will not be receptive to a new sales rep scheduling an appointment to ask them to "tell me about your restaurant" or "allow me to go through with you the portfolio of the wine I represent." They are far more receptive to an approach that focuses on their needs and interests in a way that makes efficient use of their time and attention.

The best way to approach this, to begin with, is to create a clearly defined plan. Rather than "stopping in for a chat," make a specific list of topics to be covered in the sales call, and develop sales goals for each product that you are going to present. Write these down as a quick outline that you can check during your call to make sure that you stay on topic and on target and don't miss any of your goals. This outline should include the following:

- A quick summary of your sales to the account (if an existing account) over the past period
- A comparison of these sales to those of similar accounts
- Any changes in pricing or promotional plans for the future
- Trends in the market that could have an impact on the account's business
- Suggestions for products and volumes moving forward
- Final sales goals for the call

You should arrive a few minutes early so that you can use that time to do a quick survey of the store and make sure your previous perceptions continue to be accurate. Check where your own wines are for sale and where the competition sits. (It doesn't hurt to make sure that your bottles are lined up neatly and are dusted and ready for sale!)

PREPARING TO SELL TO ORGANIZATIONAL BUYERS

As previously mentioned, selling to organizational buyers involves sales not to the ultimate guest or customer but to the organizations that serve these guests. Several dimensions distinguish the marketplace of organizational buyers from the consumer marketplace, particularly those whose purchasing power rises to the level of key accounts: derived demand, size and frequency of orders, sophistication, and buying process. These are discussed in detail next.

Derived Demand

Like your other wine-buying customers, the ultimate demand for the goods and services you are selling to the organizational buyer comes not from the organization itself but from the buying organization's customers. And in this case, the relationship is even more critical. If the buying agent of a big-box wine retailer makes a bad purchase decision, the chain may suffer financially but so also will the buying agent's personal reputation and ultimately his or her job security. To minimize such risks, organizational buyers rely on firms that have a positive reputation and are deliberate as to which distributors and sales reps they will see.

Some larger buyers are very specific about their needs and even prepare requests for proposals (RFPs) to make the sales process more transparent and minimize their risk. An airline, for example, might post a request for five thousand cases of wine at a specific price, varietal, style, or appellation and then organize a blind tasting to select the wine that best meets their goals. Cruise lines, larger hotel chains, and others might also use this kind of approach.

The smart salesperson can take several measures to reduce the risk a buyer perceives in the purchase decision. First, the salesperson will make a point of presenting a credible image—either on the phone or in person. A friendly, professional, and attentive attitude is essential, as are appropriate dress and grooming. (Buyers look for symbolism: If the salesperson is inattentive or sloppy during the "courtship" process, wouldn't he or she also be unreliable after the contract is signed?) Researching the buyer's company and having testimonials from satisfied customers in his or her industry are also valuable tools to help you build the relationship.

Size and Frequency of Orders

Organizational buyers place orders that are considerably larger than those of an individual consumer. A contract with a big-box wine specialty store may involve tens to hundreds of thousands of dollars in revenues for the distributor, who will want to negotiate both price and volume as a standard element of doing business with you. Logistics, promotional support, and quality control systems will likely come into play as well with such major key accounts. The point is you should be prepared to enter into these negotiations, either in person or via follow-up communication that allows your management team to play a role in the give-and-take negotiations with major key accounts. Patience plays a factor, too, since key accounts are generally focused on long-term buyer-supplier relationships over fixed periods. The savvy sales rep will continue to nurture the relationship during the "off years" to remain in good standing with the potential client.

The Sophistication of Organizational Buyers

Organizational buyers can be considerably more sophisticated than buyers in the consumer marketplace—at least regarding business practices. While some might not share the passionate enthusiasm of a sommelier or retail wine shop owner, they have to be very careful in considering the costs of what is being purchased. Often the organizational buying decision involves more than one individual. This panel of buyers is specifically designed to eliminate any obvious favoritism in the purchasing decisions and to give each of the members of the panel a sounding board for key decisions. The decision as to what supplier to use may also be influenced by the owner(s), management company, and franchiser. Knowing who the gatekeepers, decision influencers, and, ultimately, the decision makers are is the responsibility of the salesperson. Being versatile in presenting the right value proposition to the right decision influencer can be a critical skill when selling to a buying center.

Key Accounts Have a More Involved Buying Process

As described below, organizational buyers best characterized as national or key accounts often have standard operating procedures that guide their purchase decisions. They are as follows:

1. Recognition of needs: The potential key-account buyer outlines the needs and specifications for the wines and support services in a formal RFP.
2. Evaluation of options: The potential buyer searches for qualified potential suppliers that might satisfy the specified needs and sends them the RFP; suppliers that make the initial cut will often be invited to make a formal sales presentation to the buyer's team.
3. Resolution of concerns: The potential buyer analyzes the pros and cons of the alternative suppliers and attempts to improve the offers.
4. Purchase: The potential buyer decides among the alternative suppliers by choosing to buy from one of them.
5. Implementation: The *actual* key-account buyer becomes concerned about the integration of the purchase into his or her supply chain. He or she seeks feedback from his or her team for purposes of evaluating the supplier's performance both internally in the organization and, ultimately, how well customers are responding to the product.

Recognition of Needs

Recognition of needs means the recognition on the part of the buyer that a need exists and that a solution is available. Often it is thought that professional salespeople create needs and a call for action from a prospect. However, organizational buyers are large, sophisticated entities that seldom make major impulsive changes to their suppliers and logistical processes during a fiscal year(s). Leading key-account salespeople concentrate their time and efforts on identifying such key accounts and making sure that they are on the buyer's RFP list of potential suppliers. In addition, wise key-account salespeople will put in time up front to understand the buyer's needs and align proposed solutions their firm can provide the buyer. These efforts facilitate prompt, quick, and efficient responses to future RFPs.

Evaluation of Options

Once the specifications have been clearly outlined, the search for qualified suppliers begins. A smaller winery or distributorship with limited range in its portfolio and support services may simply not make the cut because the organization wants to make sure there is enough depth and breadth to support its needs over the long haul. A larger winery without ISO or HAACP certifications might not make the list because the organization wants a reliable supplier that has systems in place that lessen the chances of business disruption in their supply chain. These insights—often indicated in the RFP—help the salesperson build a proposal that not only covers the potential buyer's major requirements but also presents his or her products and services in the most favorable way.

Seldom is it possible to close the sale during a single call when responding to an RFP, particularly with a new account. Instead, the salesperson gathers information so he or she can prepare a proposal for the buyer. However, buyers who feel they can readily articulate their needs to a supplier often—correctly or incorrectly—feel that communicating directly with a salesperson is more of an annoyance than an asset and will not facilitate such contact at this initial stage of the buying process.

Once the buying committee has scrutinized proposals from several suppliers, they narrow their lists to finalists who then make formal face-to-face presentations. In this meeting, the sales manager or sales team will summarize their offer and respond to any concerns or special requests of the buying committee.

Resolution of Concerns

In this phase, the buyer consults with the buying team on the shortlisted set of proposals and facilitates the final decision. A good salesperson will make sure where possible that the decision influencer has all the necessary information needed to support the proposal. It is also important to remember that even after a final selection has been made, negotiations may continue regarding such matters as price, terms, schedules, and logistical and promotional support.

Purchase

The buying committee decides among the alternative proposals that will be awarded the business as either a contract or vendor agreement. The potential buyer has now been transformed into a customer. In later chapters, we will learn that the decision to purchase from one supplier and not another is based on many factors. Buyers make decisions based on the total value or benefit they will receive from a chosen supplier. And the value is not only found in the product, price, or support services themselves.

Implementation

After purchase, some customers experience what has been called buyer's remorse or postpurchase concern. If your wines or the delivery of those wines does not meet expectations, you should expect an immediate notice from large organizational buyers. Key-account salespeople know they should monitor accounts closely to ensure that everything is running smoothly. This will increase the likelihood of favorable evaluations and the probability of a long, mutually beneficial buyer-supplier relationship.

Though not all wine producers, importers, and distributors, and the sales reps that support them, can achieve the depth and range of capacity to compete successfully for national key accounts, many can. For those who believe they can, understanding this process will lead to designing more efficient and effective strategies to identify leads, receive RFPs, and make the kind of proposals that will get both the attention and the approval of significant organizational accounts.

DISCUSSION QUESTIONS

1. Consider a prospective customer you plan on calling on in the next few weeks. Based on his or her industry sector, what are at least three basic needs he or she will likely have?
2. What will you need to find out regarding his or her preferences in addressing each of these needs?
3. What products in your portfolio will be likely to appeal to him or her?

Chapter Eight

Call Opening

Opening a conversation with a prospective buyer, whether in person or over the telephone, begins with making a positive first impression followed by proposing the agenda. This is identical to the process we outlined in direct-to-consumer sales in chapter 5, but here we address it in more depth. Again, in establishing a positive first impression, you are communicating in words, tone, appearance, mannerisms, and deeds that you are credible, capable, and trustworthy. By proposing an agenda, you communicate what you would like to accomplish during the call. Let's review each step in greater detail. You never get a second chance to make a first impression.

ESTABLISHING A POSITIVE FIRST IMPRESSION

It might sound cliché to suggest in an advanced-level discussion of salesmanship that first impressions matter in the business world. We are confident that no one seriously pursuing a position in wine distribution sales needs to be reminded of the importance of good grooming and manners. Nor do we ever want to advocate that you put on airs or behave in ways that are unnatural to you. But just as the initial welcome greeting sets the tone for your customer in the tasting room or at the restaurant table, the greeting you give a trade customer can quickly establish a tone for either success or failure.

In short, how we present ourselves communicates to others in ways more powerful than words. It is this information that forms the basis of first impressions that all individuals quickly make of others, and once built, these impressions are resistant to change. The impression you make on the customer will be reflected in his or her perceptions of your company itself.

Seldom if ever would a customer want to do business with you if he or she has reason to question your credibility and trustworthiness. Though these qualities must be earned, we instinctively observe others for cues on which to base our initial judgments.

We know of no book or training program in personal selling—regardless of industry—that does not include a discussion of first impressions. In addition, there is considerable evidence in the social psychology literature that consumers look for shortcuts even in important decisions. As consultative salespeople, we want to work with our customers in a deliberative and controlled fashion to find solutions to their needs that we can satisfy and that produce win-win outcomes. However, you will find that your customers' time is tight, they can be overwhelmed by the number of suppliers asking to meet with them, and their distractions can be so intrusive that they may not always hear what you have to say even if you can gain a meeting with them. Recognizing the potential for cognitive overload, sellers who infuse their requests with one or another of the triggers described below are more likely to be successful.

THE SCIENCE OF INFLUENCE

The following discussion is derived from a noted social psychologist in persuasion, Robert Cialdini, who wrote *Influence: Science and Practice*. We strongly encourage you to read this book. Briefly described below are what he coined "weapons of influence." Though they offer no guarantee of influence, sales reps who infuse their customer interactions with these methods have often been shown to be more effective than those who do not. We contend there is nothing ethically wrong with making use of these principles if the intended outcome is beneficial for your customer.

Following are five of Cialdini's principles that we contend are particularly relevant to sales:

1. *Likability:* People are more likely to agree to offers from people they like. Several factors can influence people to like some individuals more than others, such as physical attractiveness, similarity to themselves, and paying the recipient compliments. In addition, sales reps who find ways to cooperate with customers in local philanthropic causes to achieve a common goal tend to form a bond of trust with those people.
2. *Reciprocity:* People generally feel obliged to return favors offered to them. This trait is embodied in all human cultures and is one of the human characteristics that allow us to live as a society. Offering small thank-you gifts to potential customers can at times create a need to provide something in

return. Responding to a request for charitable donations organized by an existing or potential customer may influence the recipient to reciprocate at a later date. It is never wise to make such a gift contingent on a returned favor as it may come across as a bribe or an illegal tying relationship. Instead, trust in the benevolence of ethical people to feel obligated to return favors freely given to them.

3. *Commitment and consistency:* People have a general desire to appear consistent in their behavior, and they value consistency in others. Sales managers who act on this general human desire to be consistent find ways to have a potential customer make an initial, often small, commitment. Requests can then be made that are in keeping with this initial commitment. People also have a strong desire to stand by commitments made by providing further justification and reasons for supporting them.

4. *Social proof:* People generally look to other people similar to themselves when making decisions. This is particularly noticeable in situations of uncertainty or ambiguity.

5. *Scarcity:* People tend to want things as they become less available. Items are also given a higher value when they were once in high supply but have now become scarce.

The point is that our prospective customers and buyers work in environments with ever-increasing pressures and distractions, giving them a limited amount of time to devote to your sales call. Though they—like everyone—wish to make the most thoughtful, fully considered decision possible in any situation, the accelerating pace of life frequently deprives them (and us) of the proper conditions for such a careful analysis of all the relevant pros and cons of a decision. We are increasingly forced to resort to another approach to decision making—a shortcut approach in which the decision to comply (agree, believe, or buy) is made on the basis of a single, usually reliable piece of information. The most reliable and therefore most popular triggers are likability, reciprocation, commitment, the behaviors of similar others, and scarcity.

Consider for a moment the influence principles of (1) the behavior of similar others and (2) scarcity in an experiment by the *Washington Post* in collaboration with internationally acclaimed violinist Joshua Bell. In this experiment, Bell donned a baseball cap, jeans, and T-shirt for a free forty-five-minute concert during rush hour in a Washington, D.C., subway station. Of the 1,097 people who passed by, only 7 paused to listen for a moment to his performance, and fewer still tipped the supposed street musician. No doubt, any one of these commuters would have leaped at the chance to pay to hear Bell perform at his sold-out Lincoln Center performances. However, the same virtuoso, stripped of the regalia we take as evidence of talent, produced little interest. Why? These distracted passersby instinctively prejudged Bell as a typical street musician and responded accordingly. In other words, they took what Cialdini called a shortcut and missed the opportunity of a lifetime to see a seventeenth-century Stradivarius being played by one of the world's most celebrated concert violinists only five feet away.

Cialdini recognized the ethical dilemma raised regarding the appropriateness of using the principles of influence in selling to others. The principles are ethically neutral. Their impact can be good or bad depending on the intent and benevolence of the influence agent (salesperson). According to Cialdini, salespeople can be classified into three types of influence agents: bunglers, smugglers, and sleuths. Bunglers are ineffective in using the principles and are quickly dismissed as persons who lack credibility and trustworthiness. Smugglers abuse influence principles by infusing them into influence situations in which they do not naturally reside. Sleuths are effective in using influence principles because they are more knowledgeable than bunglers and more ethical than smugglers. Salespeople who play fairly by the ethical rules in applying these principles are not the enemy; to the contrary, they are our allies in an efficient and adaptive modern economy.

As a part of our work, we sometimes organize mystery shoppers who visit stores to evaluate the salespeople and observe the customer service. We use these experiences later in sales training exercises and role plays. And we have learned that there are two elements that make an excellent first impression: a warm, friendly greeting (genuine as opposed to fake) and thanking the customer for his or her call (or for accepting your call).

In the role plays, we like to see an appropriate handshake, a smile, eye contact, and respect for a person's personal space.

PROPOSING THE AGENDA

Whenever you meet with a customer, each of you has a reason for getting together. Your reason for initiating a meeting with the customer may be to introduce yourself and your portfolio of wines, confirm or gain a better understanding of the customer's needs and wine strategy, perform a gap analysis, deliver and share samples of your wines, take an order, or any combination of the above. The customer's reason for meeting with you may include learning about your firm's portfolio of wines and support capabilities, having you educate him or her on gaps in his or her portfolio or emerging consumer trends, clarifying a proposal, or fulfilling an organizational requirement to talk with several suppliers before making a purchase decision. If your agenda and the customer's agenda are at odds, the meeting will not be very productive for either of you. Your goal is to reach an agreement with the customer on what will be covered or accomplished during the call.

As one of our trade contacts said to us, "An effective salesperson communicates the purpose of his visit, so I can make up my mind whether my time is being wasted or used wisely."

When to Propose an Agenda

You initiate a proposed agenda after your initial greeting when you and the customer are ready to get down to business. Your attempt to establish a positive first impression and thank the customer for the opportunity to speak with him or her is a first step in establishing a friendly and comfortable tone that sets the stage for an open exchange of information. After greeting the customer, you'll probably engage in some small talk to open up the channels of communication.

It's important to be sensitive to a customer's need to build or reestablish rapport at the beginning of a call. Before too long, however, you want to focus on business and the reasons for the meeting. That's when you propose the agenda.

In this agenda-setting stage, you (1) propose an agenda, (2) state the value to the customer, and (3) check for acceptance.

Proposing an Agenda

You propose an agenda by saying what you'd like to do or accomplish during the call. This sets a clear direction for your conversation and lets you establish a focus on the customer. You might say something like the following in a meeting with the corporate wine buyer for a grocery store chain:

- Your account is new to my company. Though I have spent some time analyzing your wine portfolio on your shelves, I'd like to briefly confirm if my understanding is correct so that I can offer wines that will do well for you.

Or

- I can appreciate that for your stores you need support from your distributors in merchandising your wines. What I'd like to do today is learn more about what support you expect. Specifically, what is important to you in scheduling deliveries, merchandising, and promotional support?

With a retail customer, you might begin with the following:

• Thank you for visiting today. How can I help you find a wine that you are going to really love? I'm going to find something that fits you to a T.

Or

• It's great to see you here again. How did you enjoy the last wines you bought? Are you looking for more of the same, or can I show you a few of our new releases this time?

Stating the Value to the Customer

After proposing an agenda, explain its value to the customer. This lets the customer know how the meeting will make constructive use of his or her time and further establishes your focus on his or her needs. You might say something like this:

• That way, I'll be able to perform a better gap analysis of your current inventory in order to identify wines for you to consider that will drive new sales consistent with your overall strategy.

Or

• With this information, I will be able to find an optimal solution for you that will not only conform to your needs but also do so in a way that increases sales at the best possible price to you.

Checking for Acceptance

After proposing an agenda and stating its value to the customer, you'll want to make sure the customer accepts the agenda you've proposed and doesn't have anything to add. You might use one of these questions:

• How does that sound?
• Is there something else you'd like to cover?

Checking for acceptance gives you the information you need to use your own and the customer's time productively and ensures that you and the customer move forward together. Coming prepared with a basic understanding of your customer's needs allows you to mine deeper into his or her preferences where necessary, allowing the time spent to be as brief, efficient, and productive as possible.

REQUESTING PERMISSION TO PERFORM A GAP ANALYSIS

In your precontact research stage, you will have briefly reviewed the wine portfolio of the prospect, looking for evidence of his or her wine-buying strategy. In the case of an off-premise retailer, you may also wish to review its in-store merchandising and promotional strategy. Gaining permission in the call-opening stage to perform a more in-depth analysis of the buyer's portfolio will allow you to be more precise in identifying the gaps that your company can fill.

WHAT IF THE CUSTOMER WANTS TO NEGOTIATE FIRST?

This is a tricky situation. Obviously, you never want to argue with customers and should be inclined to give them the information they ask for. However, by doing so, you are not presenting your wines or company's support services in a favorable manner. Also, you have a limited basis to offer them a specific wine at your best price—assuming finding the lowest-priced alternative to a generic varietal is their primary concern. Suppose a customer says, "Thanks, but at this juncture I am only interested in a cabernet higher than ninety points at a price lower than $120 a case." You might respond by saying something like the following:

- I certainly understand. You know what you are looking for. My portfolio has over thirty cabernets we carry in inventory at that price range. They vary in style from jammy and rich to austere and elegant. If you could spare me a few more minutes of your time to allow me to get to know what styles of cabernets you are selling and how well they are selling for you, I will be in a better position to share with you options that may be good substitutes for the poorer-selling labels you currently stock, options that will expand your range and might increase sales. Is this possible?

Or

- I certainly understand. You know what sells in your stores. However, the price I can quote you may not be our best price. I will not be able to de- termine that without knowing a few details about your merchandising and promotional support requirements. Could you spare a few minutes to help me get a better idea of what your needs are?

If the customer still shows resistance, do not argue. Simply provide him or her with the wines in your portfolio that meet his or her quality rating at the price your director of sales would expect you to provide with so many unknowns. If the inquiry comes through as an eBid or RFP, generally the customer has provided you information on the technical aspects of his or her organization's requirements. However, as you will see in the next chapter, such documents get at neither the customer's more underlying motives nor their flexibility in wines, prices, and merchandising requirements. This sharing of information is best accomplished by telephone or in a face-to-face meeting. You might respond before completing the proposal with a note saying, "Thank you for considering our wine distributorship. I will be only too pleased to fill your RFP in the hope we can earn your business. However, I have a few questions I would like to ask you first. A few minutes of your time would be helpful to me in proposing the ideal options for your company and getting you our best price. Would this be possible?"

The point is not to annoy or waste the customer's time but to discover his or her needs or "hot buttons" that may not have been fully expressed in the RFP. Asking smart, open-ended questions and attentively listening to what the customer says will allow you to build a more attractive proposal. On the other hand, leading off with a broad question such as "What are you looking for?" will demonstrate unpreparedness on the part of the sales rep, which will no doubt annoy the buyer. Put in the time to identify customer needs that are not fully articulated that may be pertinent to the desired value creation outcome. Often the experienced salesperson will tactfully offer the buyer unique perspectives that will challenge the underlying assumption in the RFP.

POSITIONING YOUR PROPOSED AGENDA

When you attempt to get down to business with a customer, it may be useful to position your proposing agenda statement—that is, to put the agenda you're about to suggest in context by relating it to other events or ideas familiar to the customer. When you've met with the customer before, a positioning statement can provide continuity by referring to your previous interaction. For example:

- In our meeting one month ago, you mentioned that you'd be cutting down on a number of fact sheets and shelf-talkers on your wine shelf to reduce clutter. I'd like to take some time to understand what effect this has had on sales.

- During our last conversation, you said you would like to learn of any locally produced wines that have a good reputation and are less than $120 a case. I'd like to bring over a chardonnay to taste and information on the vineyard.

When you are speaking with a prospective buyer for the first time, you can use a positioning statement to explain your presence and stimulate the customer's interest in talking with you. For example:

- Bob Kilpatrick (one of your store managers) suggested I talk with you about changes you are planning to make in wine inventory strategy. I'd like to understand what changes you are considering so I can select the wines in my portfolio that would make sense for you.
- Restaurant managers often tell us they need help with their wine-food pairings. I'd like to find out a little about what changes you are considering in your menu so I can have a better understanding of what wines would be right for you.

Besides putting your agenda in context, positioning statements like the above help you make a smooth transition from small talk to business talk. And remember that while every call is a sales call, customers want to feel as if you are offering more than just products for sale. One of our trade contacts told us, "It's important not to let the customer feel that you are there merely to sell your wine."

In preparing to make your proposed agenda statement, ask yourself:

- What might the customer want to accomplish by speaking with me?
- What do I want to accomplish by meeting with this customer?

Considering the customer's reasons for speaking to you ensures that you have the customer's interests in mind when you formulate your agenda. It will also help you explain the value of the agenda to the customer when you open the call. Considering your own reasons for meeting with a customer helps you set an objective for the call. With a clear objective, you can think about the ground you must cover to achieve it and formulate a focused agenda.

EXERCISE

Identify a prospective customer you will be contacting in the coming weeks. Assume that you ask for and receive permission to contact him or her by telephone for a short call. Write an opening statement you might make for that call. Remember to do the following:

- Offer a warm, friendly greeting.
- Thank him or her for taking your call or for visiting the winery.
- Propose an agenda.
- State the value to the customer.
- Check for acceptance.

If you can, write something to position your opening statement first.

DISCUSSION QUESTIONS

1. Comment on one of Cialdini's five principles and why you believe it is important.
2. List the three elements of proposing an agenda. Why is each one important?
3. What is meant by positioning your proposed agenda? Give an example of its proper use.

Chapter Nine

Probing the Customer's Needs

For you and a customer to make an informed, mutually beneficial decision, the two of you must share an understanding of the customer's needs. Your precall research helps you identify some of your customer's needs while leaving you with a vague understanding of others. Probing is the means by which you confirm these needs and gather additional information to achieve a better understanding. That's why one of our trade contacts told us, *"Good sales reps get to know my wine-buying strategy before telling me what I am missing. I have to be strategic in terms of what I stock."*

At a restaurant table, this must include finding out what kinds of wines the customer likes to drink or if he or she is celebrating an important occasion. In a tasting room, these questions help you understand why the customer is in front of you and what will make him or her happy that he or she made the decision to visit.

Probing is one of the most important skills a salesperson can develop. The ability to ask questions that logically and efficiently uncover important information about a customer's strategy and needs—and do it in a way that is comfortable and informative for the customer—is a distinguishing characteristic of the consultative salesperson. Being able to efficiently and accurately conduct a gap analysis of the customer's current wine portfolio provides you insights as to which wines in your portfolio and which merchandising support would add value to your client's range and, ultimately, sales.

Your goal in probing is to build a clear, complete, mutual understanding of a customer's needs. A clear understanding means that you can answer the following questions about each of your customers:

- What is the customer's specific wine-buying strategy?
- What aspects of this strategy are working well for the customer and what needs improving?

- What merchandising and promotional support does the customer want from suppliers?
- Based on their strategy, what are the gaps or shortcomings in the customer's portfolio in terms of wine varietals, styles, growing regions, and price points that if addressed could lead to new sales?

A complete understanding means that for the particular buying decision, you have an understanding of all the customer's needs and the priority of those needs. A mutual understanding means that you and the customer share the same knowledge and that the picture you have of the customer's needs is the same picture the customer has. A clear, complete, mutual understanding of your customer's needs will ensure that the recommendations you make to address those needs contribute to the customer's success in the most effective way possible.

WHEN TO PROBE

You probe when you wish to confirm your understanding of a customer's strategy and draw out or obtain information that you need to guide your gap analysis. The signal to probe comes from you. Whenever you feel you need more information to achieve a clear, complete, mutual understanding of a customer's strategy, needs, and circumstances, you probe. How much or how long you probe on a call depends on the complexity of the customer's needs and strategy and how clearly you understand them. Generally speaking, you should spend more time probing with a new customer if your precall research leaves you with a vague understanding of the customer's needs.

On the other hand, if you feel that your precall research gives you an adequate understanding of the customer's strategy and needs, your probing questions can be limited to confirming this understanding and what, if any, changes he or she anticipates making. Such information should be recorded into an account management system. Probing existing customers should be performed periodically to identify any changes in their needs that you should be aware of. Again, only by maintaining a clear, complete, mutual understanding of your customer's needs can you be assured that the recommendations you make to address those needs contribute to the customer's success, and in turn your success as a sales manager.

HOW TO PROBE

To build a clear, complete, mutual understanding of a customer's needs, you use open and closed probe questions to explore the customer's (1) strategy, (2) circumstances, and (3) needs.

THE CUSTOMER'S WINE STRATEGY

It is hard, sometimes impossible, to sell things. It's a lot easier to find out first what the consumer wants to buy and then help him or her make that possible. That's why a good salesperson will also ask questions before launching into a final pitch. In the tasting room or restaurant, you want to find out what wine the customer likes before you recommend a bottle. And performing a gap analysis of the customer's existing wine portfolio without first understanding the customer's wine strategy focuses on what you have to sell, not what the customer needs.

Consumers know what they like. Trade buyers generally have a good understanding of their markets and their competitive position in the marketplace in terms of understanding their best customers, what they buy, and at what price points. As such, they will see little value in being recommended wines that bend their strategy out of shape, particularly if they perceive their strategy is working well for them. By first understanding their strategy, how well it is working for them, and what they would like to change or improve, the sales rep can zero in on recommendations that fill in the gaps and adjustments in the wines they stock to serve their retail customers better.

To illustrate, a grocery store may choose to vary the wines it inventories by stocking lighter wine varietals in the summer and heartier, more full-bodied wines in the colder months. Hence, by spending the time to discover such seasonal preferences, the sales manager can make appropriate suggestions that will support this strategy. To further illustrate, a grocery store in a location close to a wine discounter may cede a degree of competitive space by not offering jug wines and low-cost premium wines because of a competitor's dominant strengths in the below-seven-dollars-per-bottle categories. Sales managers who focus their gap analysis on identifying gaps in varietals, styles, and regions at the mid-premium price levels would conform best to this customer's price strategy. Finding the customer a wine that exceeds the quality of a low-cost premium wine at a new price point of eight dollars may be of even better value to the customer.

The point is that wine retailers are highly territorial businesses, meaning that the majority of their retail customers come from a relatively small geographic market. Your customers must identify a competitive space in which they can thrive, and their competitive strategy or edge is their attempt to dominate that space. The strategy for most customers' businesses is focused on several dynamic components:

- Offering an appealing store environment and location that will attract a sufficient amount of traffic representing their targeted consumers

- Developing a reputation for providing a unique product and service mix at an attractive price that drives customer loyalty and positive word-of-mouth advertising; in the case of grocery stores, this will include their entire inventories of dry goods and perishables at competitive price points, in which the wine strategy is but a limited but complementary part
- Engaging in activities that can stimulate additional business from their customers, such as offering discounts, hosting wine tastings, matching wines with food, sharing interesting and educational wine knowledge, or publishing a newsletter
- Developing strong relationships with suppliers to help ensure the best discount deals and best supplier services obtainable
- Making timely adjustments to the above based on changing customer preferences or strategic moves by the competition

Though as a wine distribution sales manager, you may have multiple customers that overlap in a territory, your consultative goal is to assist each customer to succeed in executing his or her strategy. You will discover, and at times will have your customer share with you, information that you must keep in the strictest of confidence. Sharing proprietary information with others will undermine the trust your customers have in you and the company you represent. However, in your role of educating your customers about emerging trends in consumption, you have to share with them broad indices of consumption trends to assist them in growing their business. The line that should not be crossed is the sharing of information about a specific business or chain of businesses, whether or not they are in the same competitive set.

THE CUSTOMER'S CIRCUMSTANCES

A customer's needs do not exist in a vacuum. Customers have needs because of the circumstances that surround them. A customer's circumstances include facts, conditions, and events in his or her environment, as well as the feelings and opinions the customer has about them. In general, the more you know about the customer's circumstances, the better you'll understand his or her needs. Often, knowing the customer's circumstances helps you understand why a customer has a need. A consumer whose daughter is getting married soon might have a very different wine interest from one who lives alone and rarely entertains. To learn about a customer, you have to listen. As one of our consulting sales managers told us, "I never learn anything from my customer while I am talking."

An example of getting to know a trade customer's circumstances might go something like this:

SALESPERSON: I can appreciate that your policy is to only work with large vendors who can offer you a lot of merchandising and promotional support.

CUSTOMER: Yes, we like to operate our wine department as hands-free as possible, requiring a daily visit from a merchandiser who will restock the shelves and keep them in good order. In addition, our customers often purchase their wine on impulse, making floor stacks and end-aisle displays set up by the vendor and spread throughout the store an effective tool for us.

SALESPERSON: I can appreciate these requirements. With all the demands placed on your staff, managing deliveries, restocking wines, and finding new creative ways to market your wines in the store do not need to be added to their duties. As a relatively small wine distributor, our sole focus is providing our customers quality wines at the lowest possible prices. What we may be able to offer you in merchandising support is a commitment to maintain an ample supply in your warehouse for the one to two shelves we can supply with our labels. In addition, we will be happy to conduct wine tastings, including working with your staff on sample wine pairings for their cooking demonstrations and tastings, at times convenient to you. Given the potential of this offer to engage your customers on a personal level, could you afford the modest efforts to keep two shelves stocked by your staff?

The circumstances for most customers in business have several layers:

- Their job
- The function or department they work in
- The organization or company they work for
- The wine sales segment their organization is involved in—and the customers it serves

As a wine sales manager, you learn as much as you can about a customer's circumstances before proposing options. But when there are gaps in your understanding, or you want to find out how the customer feels about his or her circumstances, or you want to know how the customer's circumstances have changed, you get the additional information you need by probing. For example:

SALESPERSON: Are floor stacks and end-aisle displays the only effective means of increasing wine sales in your stores? Do they cause problems in your store layout and traffic flow?

CUSTOMER: Yes. If not well managed, they can also add some unnecessary clutter.

SALESPERSON: Would you allow me to get back to you with a proposal? I would like to consult with my team to come up with a range of promotional support designed to increase your sales, avoid clutter, and drive personalized service to your customers.

CUSTOMER: Sure.

SALESPERSON: That's great. Can we shoot for another meeting this time next week?

CUSTOMER: Absolutely.

Whenever you're with a customer, you listen for customer expressions of need (the language of needs). The clarity with which customers initially express their needs varies. One customer may say, "I am only interested in wines rated higher than ninety points at a price lower than $120 a case." No matter how a customer begins, you explore further until you have a clear picture of what the customer wants. To get a clearer picture of what the customer is looking for ("wines rated higher than ninety points at a price lower than $120 a case"), you might say, "What is the average inventory turnover rate for wines that meet both these criteria? Would slightly more expensive wines that meet or even exceed these turnover rates be worth considering as well? I have access to the point-of-sales data that tracks the performance of specific wines regarding price sold and average days on the shelf."

THE NEED BEHIND THE NEED

When a customer has a need, there's a reason. Sometimes the reason is another need, a need behind the one that is first expressed. The need behind the need is usually a larger goal the customer wants to accomplish and is often related to one of three general business areas: finance, performance or productivity, or image. For example:

- Expressed need: To stock only wines rated higher than ninety points at a price lower than $120 a case
- Need behind the need: To sell wines at a price point that customers will perceive as good value (performance or productivity)
- Need behind that need: To not be stuck with wines that do not sell or must be heavily discounted to clear underperforming inventory (performance or productivity)

- Expressed need: To stock wines that have sold well for us this past year
- Need behind the need: Customer may lack skills in wine pairings (finance)
- Need behind that need: Assistance in wine-food pairings so the customer can feel comfortable changing the menu (finance)

Probing to understand the "need behind the need" helps you understand why a need is important. It's particularly useful when customers are very specific about what they're looking for ("wines rated higher than ninety points at a price lower than $120 a case"), but it's not clear how you will be helping them if you provide what they want. You might ask, "What do you expect from such wines in terms of inventory turnover and net profits?" or "How confident are you that wines that meet these criteria will sell quickly for you at the price you expect?"

If the "need behind the need" is unclear, it warrants a follow-up question to confirm both its existence and the fact that you and the customer have a mutually shared understanding of it. If you believe the customer who states a need to "carry only those wines that sold well for us the previous year," you might assume that he or she may be unaware that the taste of a wine produced by the same vineyard may vary from year to year given fluctuations of temperature, rain, and grape availability. The customer may also lack skills in wine pairings. In such a case it is unwise to simply state, "I think you need to be interested in new products, and I'd like to show you my portfolio anyway." Instead, the astute sales manager would be wise to probe deeper, looking for more opportunities he or she can address. For instance, he or she could ask a follow-up question such as "May I ask, have straight rebuys of the same wines ever presented problems for you?" This approach will be the focus in chapter 11 on closing, but suffice it to say here that the astute sales rep, in a nonargumentative way, is consultatively probing for opportunities to challenge customers' status quo, helping them discover possible needs you can support in helping them grow their business.

OPEN VERSUS CLOSED PROBES

A probe is a question or request for information. One of the sales managers we consulted with suggests, "Effective salespeople ask good probing questions, ask the right follow-up questions, and ask the kinds of questions that will get customers to talk." There are two general types of probes: open probes and closed probes. Both have their appropriate uses in a sales call.

Open probes encourage customers to respond freely. For example:

- What are you looking for?
- Why is that?
- Tell me how you do that now.

Closed probes limit a customer's response to the following:

- Yes or no
- A choice between alternatives that you supply
- A single, often quantifiable, fact

For example:

- Are you aware that all orders are delivered within forty-eight hours?
- Would you consider wines that cost more than $120 per case but average quicker turnover than you currently achieve?
- If I can provide you regularly scheduled wine tastings in your store, will you be willing to absorb increased labor costs on restocking our wines on your shelves?

In general, it's a good idea to keep your probes as open as possible. Open probe questions encourage customers to speak freely and allow them to share information that they think will be useful to you. Of course, if you rely exclusively on open probes, your discussion may lack focus and may not be an efficient use of time.

Closed probes bring focus or closure to a discussion and are particularly useful when a customer rambles or is not clear on his or her commitments to you. But if you rely too heavily on closed probes, the customer may feel as if he or she is being interrogated and become unwilling to share information.

On a sales call, you use open probes in several ways:

- To gather information about a customer's circumstances: When probing to confirm or explore more deeply the facts, conditions, or events in the customer's environment, you might say, "My initial analysis of your wine portfolio suggests to me that you prefer wines from our region at the premium and luxury levels. Is this correct?" or "How are we positioned pricewise among your other vendors?"
- To uncover needs: When a customer expresses needs he or she wants to discuss with you, the best way to learn about those needs is simply to invite the customer to tell you about them. For example: "What is your preferred

method of delivery?" or "What kinds of help are you looking for from your wine vendors?" or "What problems do you want to avoid in working with your new wine vendors?" Once you've clearly understood one need (and provided information about how you can satisfy it), continue probing. You might ask: "What else will be important in adding a new wine vendor?" or "What other kinds of help are you looking for?" or "What problems have you experienced in the past that you would like to avoid in managing your wine department?"

GAP ANALYSIS FOR THE TRADE

Once you have a clear understanding of the buyer's strategy, circumstances, and needs, it is time to perform a gap analysis of his or her wine portfolio (figure 9.1). As mentioned in chapter 6, a gap analysis is a method used by wine distribution sales reps to identify gaps in the range and depth of a customer's wine portfolio that a supplier may fill or improve. A gap analysis, as currently practiced in the wine industry, can take one of two basic forms. A product-focused form consists of (1) a template that highlights the varietals, styles, countries of origin, and price points of wines in a supplier's portfolio; (2) a check sheet used to determine the depth and breadth of the customer's current wine inventory; and (3) a means to summarize where the gaps exist that need to be filled. The original template or check sheet is based on the wine supplier's wine portfolio in order to assess whether the supplier can add additional value to the customer's portfolio and how.

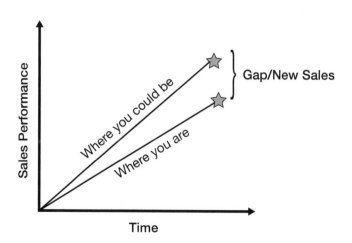

Figure 9.1. Helping Your Customer with a Gap Analysis

An alternative sales-focused approach involves identifying missing wines in a customer's portfolio that are selling well in other similar settings. The logic here is that there are just too many wine varietals, styles, and vintages to stock. Filling the gaps with wines that are currently selling well in other locations adds value to a buyer focused on increasing sales and inventory turnover by capitalizing on changes to consumer preferences. Which approach to use should be based on your understanding of the customer's strategy, needs, and circumstances. For example, a customer in a wine specialty store or fine-dining restaurant may see little advantage in stocking wines that most of his competitors stock and sell in large amounts. Instead, he is looking for the new and different in an effort to drive new trends.

We contend that the presentation of the gaps to some customers could be further strengthened with a hybrid of the two approaches. Recommending wines that are (1) aligned with the customer's wine-buying strategy and needs (determined in the precall research and initial probing stages) and (2) supported with evidence of how well each recommended wine is selling across the segment, thus supporting the competitive niche that the buyer is attempting to exploit and dominate. To illustrate, wine specialty store owners have often shared that though they are always interested in adding wines to their shelves, they do not have the time or interest to go through a sales manager's portfolio or to spend a lot of time with the salesperson explaining their wine-buying strategy or needs. They want a sales rep to make their lives easier by presenting them with options that add depth or range to their inventory strategy as well as information on how well they sell.

In addition, we contend that to further enhance the consultative nature of the sales call for off-premise retails, the gap analysis can and should include an assessment of customers' merchandising and in-store promotion activities to show deficiencies that should be addressed (figure 9.2). By doing this, the sales rep will be demonstrating where the customer currently is with his or her wine portfolio, contrasted with where he or she could be by filling in the gaps and making adjustments to the portfolio in terms of new sales.

Obviously, it is impossible to provide guarantees about what the recommended changes to a customer's portfolio would mean in terms of new sales. However, presenting him or her with a system that takes into account the recent sales of proposed wines across similar sectors is a basis for a credible case. Most sales managers naturally provide such case-bound evidence by showing that each wine has been selling well in similar locations. Doing so while drawing from a regional sales database, we contend, would provide stronger, more convincing evidence that the proposed changes in the customer's portfolio will drive new sales. Such an approach also offers insight

	Time	Latest 12 Weeks Ending 10-07-18
	Geography	STATE OF WASHINGTON - MULO + C (W/B)
	Price Segment	(All)
	Domestic vs. Imported	(All)
	Place of Origin	(All)
	Size Package	(All)
	Supplier	(All)
	Brand	(All)
	Category	(All)

Rank	Flavor	$ Share	$ Share Chg	$ Sales	$ Sales Chg	$ Sales % Chg	Eq Vol Share
1	CHARDONNAY	18.3%	(0.6)	$20,569,458	($508,014)	(2.4%)	(20.4%)
2	CABERNET SAUVIGNON	13.1%	(0.3)	$14,734,495	($145,967)	(1.0%)	(11.6%)
3	PINOT GRIGIO/PINOT GRIS	9.6%	0.4	$10,762,889	$547,196	5.4%	10.6%
4	ALL OTHER RED BLENDS	9.1%	(0.1)	$10,274,512	($54,105)	(0.5%)	(7.6%)
5	FUME/SAUVIGNON BLANC	5.5%	0.0	$6,191,800	$81,841	1.3%	4.4%
6	MERLOT	5.1%	(0.4)	$5,704,670	($385,466)	(6.3%)	(6.1%)
7	ALL OTHER BLUSH VARIETALS	3.8%	1.3	$4,330,655	$1,468,947	51.3%	4.5%
8	PINOT NOIR	3.4%	0.0	$3,857,333	$73,788	2.0%	2.4%
9	ALL OTHER BLUSH VARIETALS	2.9%	0.2	$3,245,049	$261,187	8.8%	1.8%
10	BRUT SPARKLING WINE	2.4%	0.0	$2,715,758	$77,764	2.9%	1.8%
11	ALL OTHER RED WINE NON VARIETAL	2.3%	0.4	$2,580,171	$504,659	24.3%	2.5%
12	RIESLING	2.3%	(0.3)	$2,538,040	($285,754)	(10.1%)	(2.4%)
13	PROSECCO SPARKLING WINE	2.2%	0.1	$2,479,063	$176,657	7.7%	1.2%

Figure 9.2. Gap Analysis Identifying Missing Wines in a Customer's Portfolio That Are Selling Well in Similar Settings

into your customers. As one sales manager told us, "As long as the customers know there is some method to my madness, they'll be a lot more open."

Since templates are based on each supplier's current portfolio, they are proprietary in nature. Hence, it is beyond the scope of this book to share specific templates and check-sheet examples used in the wine industry. However, the technique of performing a gap analysis is well developed in the business literature and by following the steps outlined above would be relatively easy to construct. Furthermore, a number of vendors produce smartphone apps customized for each client that can make the process quick and efficient.

IN CONCLUSION

Consultative selling can be thought of as needs-based selling where first you attempt to understand the customer's needs and only then attempt to sell your product. With good probing questions, customers will often share information on what they are looking for and what is important to them in making wise wine-buying decisions. Gap analysis further strengthens a sales rep's understanding of a customer's needs by discovering deficiencies in his or her wine portfolio and merchandising efforts that may be keeping sales at suboptimal levels. If you do this, you will be thinking like your customers. And as one of our sales managers told us, "Good salespeople should know how to sell things by putting themselves in the customer's shoes."

DISCUSSION QUESTIONS

1. Identify a gap in a customer's portfolio that you recently identified. What specifically did you find missing?
2. Did you convince the customer that the gap was important? Did it lead to a sale?
3. Were the wine's price point or attributes aligned with the customer's wine-buying strategy? Did you provide evidence as to the ability of the new wine to generate sales?
4. What would you do differently?

Chapter Ten

Supporting the
Needs of Your Customers

Your use of probing questions is the means by which you gather the information you need to ensure a clear, complete, mutual understanding of a customer's strategy and needs. To help the customer make an informed buying decision, you must also conduct a gap analysis that provides information about the wines and support services with which you and your company can address those needs. You want customers to know how you can help them, and they want to know as well.

For consumers, this can often be as simple as agreeing with their opinion of your wines or citing the awards or critical acclaim those wines have won. But it also involves sharing a personal story about how you made the same decision to purchase that wine and the success and enjoyment you gained by doing so. This lets consumers know you understand them, hear them, and can speak to them via your own experience.

Supporting is the skill you use to provide information about your products and services at the time and in a way that's most meaningful and compelling to the customer. Your goal in supporting is to help a customer understand the specific ways in which your products and support services can satisfy a need that he or she has expressed or identified in the gap analysis.

PROVIDING INFORMATION
TO ADDRESS CUSTOMER NEEDS

When you are in a one-on-one meeting with your customer, you support by demonstrating your company's abilities to satisfy each of his or her identified needs through matching statements and proof devices.

Figure 10.1.   Succeed by Helping Your Customer Succeed

HOW TO SUPPORT

Supporting statements match the potential customer's needs with your company's capabilities through verbal statements with appropriate proof devices. This is where you present features of your company's wines and support services as benefits. In our mystery shopping exercises, we find that salespeople too often merely state a feature of their products or services without translating its benefit to the customer. In effect, they are assuming that customers will see the obvious connection between the product's feature and how it will directly benefit them.

Recall what we spent time discussing in chapter 3, that customers do not buy things—they buy value and benefits. So a salesperson should never leave it up to chance that the customer will see the benefits. This is why salespeople must use fully developed supporting statements.

PRESENTING FEATURES AS BENEFITS

When you support, you provide information about your wines or merchandising and promotional services. There are two ways you can do this—in terms of features and in terms of benefits:

Table 10.1. Translating Features into Benefits

Feature	Benefit
A wine rated with a ninety score or higher	A wine that your consumers who use Vivino or other wine apps will consider to be of good quality
A wine organically grown and bottled	A wine that will appeal to your health-conscious consumers
Three-day-a-week delivery schedules	With appropriate planning and forecasting, assurance that wine inventories will remain appropriately stocked
Wine we stock in such short supply that it is offered exclusively to fine-dining restaurants	Your customers will not find the wine offered at retailers for less money

- Feature: A characteristic of your wines and your company's support services
- Benefit: What a feature means to a customer

Generally speaking, a feature of a wine is some fact about the wine. A benefit is the value of that feature to the customer—the particular way in which the feature addresses the customer's need and/or improves the customer's circumstances. Here are some examples of features and related benefits:

- Feature #1: A wine rated with a score of ninety or higher
- Benefit #1: A wine that most of the buyer's customers will perceive to be of good quality
- Feature #2: A wine that is organically grown and bottled
- Benefit #2: No chemical residuals or sulfates that might concern the buyer's customers
- Feature #3: Wines we stock in limited supplies that are offered exclusively to fine-dining restaurants
- Benefit #3: Customers will not find the wine offered at local retailers for less money
- Feature #4: Three-day-a-week delivery schedules
- Benefit #4: Even in unexpectedly high demand periods, the customer will seldom run low on inventory

If you talk about the features of what you sell without describing corresponding benefits, the customer may not understand how the feature you've described addresses his or her need, and his or her reaction may be "So what?"

By describing benefits, you link the features of your product and organization to the customer's needs.

PROFILES OF FEATURES AND BENEFITS

As a sales manager prepares and gains experience with his or her products and services, he or she will often create profiles of the features they offer and the benefits those features can deliver. Over time, the profile list will grow as you discover needs that your product or service can satisfy. By writing them down, you will commit them to memory and can respond at the appropriate time. In the next section are some of the features and benefits a sales manager of a wine distributor might construct. Note that a single feature often has more than one benefit.

ADDING PROOF DEVICES
TO PROFILES OF FEATURES AND BENEFITS

It is said that a picture is worth a thousand words and a demonstration is worth a thousand pictures. A lot can be riding on your customer's ability to make a wise purchase decision. So having available, at the appropriate time, indisputable proof to support your matching statements will make your presentation more compelling. Building on the features and benefits above, a salesperson might say:

- Here is a fact sheet on this cabernet I believe will work well for you. As you can see, it has consistently scored above ninety year after year, which attests to the winery's focus on quality. What this means to you is that you can be assured that no matter what source your customers use to gauge the quality of your wine, this cab will be on the top of the list.
- I understand that your ability to store your inventory of wine is limited. Here is our delivery schedule, which shows that our trucks pass by your restaurant Monday, Wednesday, and Friday mornings every week. With a twelve-hour advance notice and with proper inventory controls, you should never run out of any of our wines on your wine list.

A proof device, when presented well, reinforces the matching statement where features are translated into benefits. It can be as simple as a fact sheet or brochure on a wine or visuals of the merchandising collateral that you can provide. Testimonial or thank-you letters solicited from satisfied customers commenting on a commonly valued feature of your products and services can also be used as proof devices in support of a specific supporting statement. However, proof devices can also take the form of wine tastings and invitations to the customer to tour your facility and/or to tour the vineyard whose wines they sell.

It is the responsibility of all sales managers in their precall preparations to have a clear understanding of their product or service's features and potential benefits and source the appropriate proof devices that will reinforce one's matching statements. Understandably, this is easier when you have multiple contacts with the customer where (1) your first meeting is designed to gain a clear and complete understanding of the customer needs, including the gap analysis, and (2) your second meeting is when you return with a proposal. Often, having multiple contacts is the norm with buyers who employ buying committees in reaching a decision. However, a sales manager should never be surprised by a customer who desires to compress the process into a single contact. In such circumstances, anticipation is critical so that you react to each need discovered with an appropriate matching statement and proof device.

PRESENTING FEATURES AS BENEFITS

Identify two features of the wines and support services you offer (or one day would like to offer). For each feature, describe one benefit—what that feature might mean to a customer. Next, identify appropriate proof devices that can reinforce these capabilities.

Feature   Benefits

_____   _____

_____   _____

Proof Device: _____

Feature   Benefits

_____   _____

_____   _____

Proof Device: _____

WHEN TO SUPPORT

It's appropriate to use the skill of supporting when (1) the customer has expressed a need, (2) you both clearly understand the need, and (3) you know how your organization can address the need. If the customer has not expressed a need but allowed you to perform a gap analysis, you support when (1) you have identified a gap in the customer's portfolio that you believe is consistent with his or her wine-buying strategy, (2) you know how your organization can address the deficiency, and (3) you are prepared to produce evidence to support the potential benefits of addressing this deficiency. If you make a supporting statement before all three of these conditions are met, you run the following risks:

- The customer may feel that you're interested only in pushing your wines and may doubt your commitment to providing him or her solutions that will lead to increased sales (e.g., sales lift as opposed to sales shifts).
- The information you provide may not be as specific or as helpful as it could be.
- The wines you talk about may not be the ones that best address the customer's need.

In general, it's a good idea to make a supporting statement for each need separately, after you and the customer have discussed it. That way, you provide information at a time when it's meaningful to the customer, and the exchange of information between you proceeds as a dialogue. Of course, there may be times when you're ready to make a supporting statement but the customer isn't yet ready to listen. If you sense that a customer wants to describe additional needs before finding out how you can help, you can wait and support several needs together.

HOW TO SUPPORT A NEED

A feature presented as a benefit that is not linked to a customer's need is irrelevant. Presenting those runs the risk that your customer will see you as more interested in selling him or her something than in helping him or her solve a problem or accomplish what is important to him or her. Therefore you must link your matching statements to your customer's needs.

To link your product or service to a need you have identified either by probing or gap analysis, you (1) acknowledge the need or gap, (2) describe the relevant features and benefits, (3) offer relevant proof devices, and (4) check for acceptance.

ACKNOWLEDGING THE NEED

One way to promote an open exchange between you and the customer is to acknowledge the customer's needs—that is, to show that you understand and respect his or her needs. One trade customer told us, "The capacity to listen and demonstrate understanding is quite important." A good time to do this is when you make the transition from questioning to supporting.

You might say:

- I can understand why this is important.
- I can understand why you need to be strategic in the wines you inventory. It makes little sense to cover a wide range of wines available if they all will not sell quickly without discounting.

Acknowledging creates a sense of mutually shared understanding or harmony between you and the customer. It says, in effect, "I'm with you. I understand your point of view, I respect your priorities, I empathize with your feelings, and I support your desire to accomplish this task." It prepares the customer to hear what you have to offer and encourages him or her to share additional needs. And one of those needs is expertise. One sales manager told us, "My customers don't have time to become familiar with all the wines I sell, so they rely on me to have the expertise."

WAYS TO ACKNOWLEDGE NEEDS

There are many ways to acknowledge needs. For example:

1. You can agree that the need is worth addressing.
   - That makes sense.
   - I think you're right to keep your house wines at a certain price point.
   - That is important for a retail shop like yours.
2. You can mention the importance of the need to others.
   - Many of the customers we work with have shared similar needs.
   - I talk to a lot of restaurant owners who have reached the same conclusion.
   - You're not alone in that viewpoint.

3. You can show that you recognize the consequences of not satisfying the need.
   • Right. If you cannot receive frequent deliveries, you run the risk of losing sales and disappointing your customers.
   • Definitely. You don't want your waitstaff to have to apologize for not having wines on your wine list.
4. You can demonstrate your awareness of the feelings that surround the need.
   • That must have been very frustrating.
   • Sounds like a very challenging mandate.
   • It's expensive to have to print out last-minute changes to your wine list.

It's not essential to be able to recognize or categorize different types of acknowledgments. It *is* important to become comfortable acknowledging in a variety of ways, so you're not always saying, and the customer doesn't always hear, "I understand."

OTHER TIMES TO ACKNOWLEDGE

Acknowledging is useful as the first step of supporting because it prepares the customer to listen and helps you make a smooth transition from the exploration of a need to an explanation of how you can help. There are other times when acknowledging is useful:

• If a customer expresses a need but you're not ready to support it (e.g., because the customer isn't yet ready to listen or because you're not yet sure how you can help), you can demonstrate your understanding and respect by acknowledging the need anyway.
• Besides acknowledging needs, you can acknowledge the information a customer provides ("That's quite useful to know") as well as the feelings or opinions he or she expresses ("I appreciate your concern" or "That's an interesting way of looking at it").

Of course, you don't want to acknowledge everything a customer says. To be effective, your acknowledging statements must be sincere and reflect genuine empathy, understanding, or respect on your part.

1. Acknowledge the need.
2. Describe the relevant features and benefits.
3. Offer relevant proof devices.
4. Check for acceptance.

DESCRIBING RELEVANT FEATURES AND BENEFITS

When describing features and benefits in a supporting statement, it doesn't matter whether you start with the feature or start with the benefit. You might say:

- We carry over five thousand wine labels covering all varietals and styles in our inventory. That means we can find the ideal wines to pair with your menu at a price and quality that works for you. (feature first)
- We can find the wines that pair well with your menu at the price and quality you expect. That's because our distribution center stocks over five thousand wine labels covering all varietals. (benefit first)

What does matter is that the features and benefits you describe are relevant. After all, your product and organization have many features and even more benefits. When you make a supporting or matching statement, you want to describe only those features and benefits that address the particular need you are supporting. For example:

- *Need:* It's important that the wines on my wine list are not available at retailers or midlevel restaurants.
- *Feature:* My company has a range of premium wines we can only purchase in limited quantities. Big-box wine and grocery stores are not interested in wines of such limited quantities. As a result, we make these available only to fine-dining restaurants like yours. Once our inventory is contracted for, no other business in this market will be able to source it.
- *Relevant benefit:* You can be assured that once you place an order, you will be the only restaurant in this region that can offer this wine to your guests, so if they enjoy the wine they will learn there is no other place to find it.

LINKING BENEFITS TO THE NEED BEHIND THE NEED

If you've done an excellent job of probing to understand why a need is important, you may have discovered a need behind the need initially expressed by the customer. If so, you can strengthen your supporting statement by linking the benefits you describe to the need behind the need that you and the customer have discussed. A trade customer told us, "A good salesperson can answer the question, 'If we decide to do business with you, how will my business benefit?' I always ask sales managers this question, but I rarely get an answer."

For example:

- *Need:* I need assistance with wine pairings.
- *Need behind the need:* Sales have been lagging, requiring significant changes to the menu.
- *Support statement:* My sales director is a former chef himself, having trained at the Cordon Bleu in San Francisco. He will be happy to work hand in glove with your chef and wine stewards to find the perfect pairings for your new menu. In addition, I have access to point-of-sales data on each of the wines in our inventory. That means I can give you specific information on each wine's popularity so you can evaluate the profitability of each option. We want your revised menu to be a success. That way we can earn your business in the future.

OFFER RELEVANT PROOF DEVICES

Again, this is where you provide the customer evidence that demonstrates your ability to satisfy the customer's need with the stated feature and benefit. Proof devices are tangible illustrations that support and complement your supporting statements.

Should all supporting statements be matched with a relevant proof device? No. You can quickly overwhelm and distract a customer by offering too much information to absorb. Use them sensibly where they uniquely support the critical need of the customer. In chapter 12 we will discuss situations where the customer has shared doubts or misunderstandings about your capabilities. Proof devices here will be instrumental in working through them.

CHECK FOR ACCEPTANCE

After describing relevant features and benefits and offering the relevant proof devices, it's time to check the customer's reaction. You don't want to move ahead until you know that your explanation was understood and the benefits you described have been accepted.

In checking for acceptance, keep in mind that you don't have to check verbally. If this is a face-to-face meeting, usually it's sufficient to make eye contact with the customer and assess his or her reaction to the information you've provided and respond accordingly.

If there's any indication that a customer doesn't understand or accept the benefits you mention, probe to find out what's on the customer's mind and handle the confusion or concern right away. If you cannot tell whether a customer accepts the benefits you introduce, ask a question such as "How does that sound?" or "Would that work for you?" If the customer reacts favorably to a benefit, make a mental or written note. When it's time to close, you'll want to remember which benefits were accepted by the customer.

IN CONCLUSION

Supporting is an essential element of the consultative selling process. It is where you demonstrate to your customer the value of your wines and support services. However, you cannot make an effective supporting statement until you understand a customer's unique need—what he or she wants and why it's important. The best way to prepare yourself to make the connection effectively is to do the following:

1. Know thoroughly the features and benefits of what you sell.
2. Prior to making a sales call, think about the range of needs your customers will likely have and how you might address them. Assemble and organize your proof devices. Planning is the key.
3. Before making a supporting statement during a call, ask yourself:
   a. Has the customer actually expressed a need? If not, can you identify needs in a gap analysis?
   b. Do I know how my organization can satisfy the need or fill the gap? If not, acknowledge the need, but don't make a complete supporting statement until you can describe relevant features and benefits.

Once you can answer yes to these questions, it's time to support. Go back to our outline:

1. Acknowledge the need.
2. Describe the relevant features and benefits.
3. Offer relevant proof devices.
4. Check for acceptance.

Supporting—like all elements of the sales call—should come across as a conversation you have with the customer. No doubt you will feel awkward the first few times you do it, either in training role plays or in front of a customer, but with the effort, you will become more comfortable.

THE ART AND SCIENCE OF STORYTELLING

In the wine business, what we communicate and the way we communicate are all too often ineffective. We have all sat through presentations that endlessly provide every detail and fact about a winery and a wine, and yet we are still left feeling as if it has been more of a mindless recitation out of the encyclopedia than a real attempt to communicate and make human contact. We are not excited. We are not touched. We are not moved.

The *human* contact is what sells wine.

The solution to this challenge is to spend less time reciting the facts and more time creating the kind of experience that your listeners will remember. Do you think they are going to remember all those facts? Almost certainly not. Most teachers will tell you that their students might remember two or three critical facts from a lecture, but they will never remember a long list of them. Not even winemakers can remember all the details of wines they produced with their own hands, so it is unlikely that your listeners will remember them either. And why should they?

The situation gets worse when you understand the communication challenges of the three-tier system. Passing on a litany of soil types, farming data, harvest dates, Brix levels, acid measurements, barrel regimens, and the rest will make anyone's eyes glaze over. And we certainly can't count on those listeners to turn around and repeat the same information through all the levels of the distribution system, from winemaker to distributor to restaurateur to sommelier to consumer. It's like a version of the old parlor game of the broken telephone, where a message is whispered from ear to ear and ends in chaos and silliness. What gets through a system like this? Something really memorable. And while we have already established that most facts alone are not memorable, a good story always is.

If you tell a good enough story, the story will get repeated through that same system and beyond, because people love a good story. Your job as a salesperson is to develop the kinds of stories that people like to hear and like to tell. As the old saying goes, facts tell but stories sell.*

So what is a good story? The first criterion is that stories have to be about people. There has never been a good story about an inanimate object. What makes the story great is how people react. Even a front-page news story about the latest scientific invention that will transform your life quickly moves from the invention itself to the story of the man or woman who invented it and the story of someone whose life was transformed by it. Make your stories about people, not things. We relate to people. We don't relate to things.

---

* Paul Wagner first heard the phrase "facts tell but stories sell" from his old friend Sharon McCarthy of Banfi Wines many years ago, and he has used it ever since.

Stories also have a plot. A plot is something that allows us to see how the people react. It teaches us something about the people in the story, and it allows us to imagine ourselves in the same situation. Suddenly, instead of being bored by a catalog of numbers and dates, we are involved in what is being said to us. We are in direct human contact with the storyteller—and we are touched. Below are described the various parts of a story and how they function.

You begin with the *exposition*, where you set the stage for what comes next. In the exposition, you introduce the characters and a bit about their relationship. For example, you could explain that the winery you're introducing is family run and that each member of the family has a role to play, or that their wines are made with grapes from specific vineyards, so the winemaker and vineyard owner need to work together to achieve success. You set the foundation for the story with this simple exposition, but if your story doesn't get beyond the explanation of the characters, you don't have a story, and you won't be successful.

Once you've set the characters with the exposition, the next step is to bring them to life with an *inciting incident*. Stories need tension to be interesting. If there is no problem, no tension in the situation, then the story has no future. Maybe all the family members don't agree on every decision? Instead of hiding this situation behind the walls of the winery, explore how the tension forces everyone in the family to pay more attention, to work harder, to make better wine. To continue this example of a husband and wife team, you might include something like "We make our rosé because my husband really likes rosé. And even though I was not as excited about rosé as he is, I decided that if we were going to make a rosé, it was going to be a wine that I could really love." With an introduction like that, you will have people smiling and paying attention to your explanation of how and why they make this wine.

Or, taking the other person's point of view for your inciting incident: "I always loved the grapes I got from the blocks in the middle of John's vineyard and wanted to make a wine that showed those off. But he told me that everyone else likes those same blocks, and he wasn't excited about selling all those grapes to me . . ." Okay, you've got our attention. How did the speaker convince John to sell him his best vineyard blocks?

What follows the initial crisis of the inciting incident is the *action* of the story. Be careful here—what may seem like a fascinating series of events and developments to you may end up being too long and involved for a good story. You want your listeners to remember and retell this story. Don't make it so long and complicated that they can't or don't want to do that. A little development goes a long way, especially in a sales presentation. When you have time over dinner to tell the whole story, you can indulge in a little more detail. We have all sat through a story that was too long and wished the teller would cut to the chase. We know how that feels, so edit ruthlessly. Make a long story short. If a detail doesn't add drama to the story, it isn't necessary. When in doubt, leave it out.

As the action develops, your listeners will want to know what finally happened, and particularly as part of a sales presentation, you shouldn't make them wait too long. They have other things to do!

Which brings us to the *climax*—the key part of the story that is the most interesting and reveals the most about the characters. In a joke, it is the punch line and should be delivered with enough energy to make it hit home with your listeners. "So I told my husband that I wanted to make a rosé that would make him forget about all the other rosés in the world. And that's what you have in the glass in front of you." That makes you want to try the wine, doesn't it? Notice that she has not said a thing about how she makes the wine—not a word about viticulture or enology. And yet she's got you ready to like the wine. That's good salesmanship. Or, speaking from another point of view: "I got tired of negotiating with John every year about the same thing and never getting quite what I wanted. So I went behind his back and married his daughter." And of course, he now gets exactly the vineyard blocks he wants. As a result of the story, you not only want to taste the wine but you've also developed an appreciation for the winemaker and his sense of humor. You've made human contact, and you associate it with the wine.

There is one final element to any good story, which is that after the climax there should be some final note of *resolution*. This is where you tie up loose ends a bit and let people know that this story is now over. It also allows the listener to relax a bit after the tension of the story. "My husband absolutely loves my rosé, and it is his favorite. But I still catch him looking at other rosés from time to time. He just better not let me catch him drinking one." That line should get a laugh and make the human connection that will help you build a long-term relationship with your customer. Another approach: "So I got the grapes I wanted and the wife I wanted. But now I have to negotiate with my wife about what kinds of wines I make . . . and she learned a lot from her dad." That's a particularly useful line because now you've already set up the exposition and inciting incident to tell more stories. You are on a roll!

Pixar writer and director Andrew Stanton has five rules for a good story that he shared in a TED Talk. Here are his bullet points, with a note of explanation from us, based on his comments:

- *Make me care:* Be a sympathetic person I care about. Admitting a negative about yourself often generates a very positive image in your listeners. If you admit that you weren't a fan of rosé, you already may get some sympathy from your customers. If you admit that you now have to negotiate with your wife, we want to know how that is going!

- *Take me with you:* Invite us on the quest. So now we have the problem to solve: How are we going to make that rosé in a way that will be great—that will make her husband forget all other rosés? You could tell a little bit about your trials and errors because we want to share your quest. But not too long!
- *Be intentional:* Explain the motivation. When our winemaker insists that he needs those specific blocks because he wants to make a great wine, not just a good wine, we gain an understanding of who he is and what drives him. And we look for other ways to expose that same trait in him.
- *Let me like you:* Create empathy. As you struggle to achieve your goal, the listener will begin to share your feelings and experiences. That is the essence of creating a human bond with your audience.
- *Delight me:* Charm and fascinate. After you have invited me to share your emotions as you struggled through your challenges, you should also invite me to share in your success. You have taken your listeners on a journey, and they are delighted with the final destination.

By putting all these elements into your stories, you will become a more effective salesperson. But even a short anecdote can incorporate some of this same thinking into a response that will generate the same reaction in your listeners. Some good salespeople do this instinctively, but for the rest of us, it takes awareness and practice.

When you are asked if a winery owner believes in organic farming, you might answer yes. But then the questions come hard and fast: What system of certification does he use? Does he agree with this specific technique? How does he manage this specific problem? You could spend quite a long time answering each of these questions in great detail if you know the answers and the details. But a good storyteller might take an alternative approach: "Look, I don't know all the ins and outs of his certification, but I know that he and his wife let their three little children sleep out in a tent in the vineyard all summer long. The kids live out there, and his wife told me that they farm organically because they couldn't imagine doing it any other way." That's a solid, human answer to a series of very technical questions, and it does an outstanding job of communicating the winery's dedication to organic farming. And that story is much more likely to get repeated through the system than a long catechism of organic growing standards.

Finally, a note about what kinds of stories work best. In the wine business, many people like to cite esoteric sources and arcane data as a way of making us understand that they are real experts. Storytelling doesn't work this way.

Yes, you can include the name of a top master sommelier as a reference in a story to someone who thinks that master sommelier is a true visionary. In fact, that's a good use of your industry knowledge. But don't confuse reciting facts and dropping names with telling a good story.

A good story will hold up just fine when you tell it to a smart fifth-grader. If you can keep that child interested in the story until the end, it will work perfectly well with adults. If that child starts to lose eye contact and look bored, you'd better do a little work on the story. Edit it, cut to the chase, and make sure that it works from beginning to end.

DISCUSSION QUESTIONS

1. How should features, benefits, and proof devices be linked to customer needs?
2. How do you support a need?
3. Consider for a moment your job searches after graduation. Describe two unique features (skills, experience) you have and their potential benefits to an employer. Recall that a feature can have multiple benefits. In addition, describe a proof device that you could use to reinforce each of these matching statements.

   Feature #1: Benefit

   Proof Device: _____

   Feature #2: Benefit

   Proof Device: _____

Chapter Eleven

Closing the Sale

You build the foundation for an informed, mutually beneficial decision by understanding the customer's needs and helping the customer understand how your product and service can address those needs. Often during this exchange of information, the customer will share concerns that may take the form of skepticism, false impressions, and shortcomings (the topic of chapter 12). If these are managed well, both you and the customer will be ready to move ahead, and closing will be a natural next step in this dialogue.

Your goal in closing is to reach an agreement with the customer on the appropriate next steps, if any, for moving a mutually beneficial decision forward. You use the skill of closing to advance the sales call to its next level of progression as well as to close sales. Whether this be in a trade situation with a major client or at the sales counter of a winery tasting room, an effective close can increase sales and lay the groundwork for a much longer relationship. As one sales manager told us, "The customer sends out signals. It is important to be very sensitive to these signals, not to go on to discuss other things and miss the key moment to ask for a commitment."

WHEN TO CLOSE

You use the skill of closing when (1) the customer signals a readiness to move ahead ("buying" signal) or (2) the customer has accepted the benefits you've described. You can feel comfortable asking for his or her business if the customer gives you a *verbal* or *nonverbal* signal that he or she is interested in moving forward. For example:

- It all sounds good.
- Include in your proposal . . .
- I like what I am hearing.
- Smiling, nodding, or looking at you with interest.

If a customer does not signal a readiness to move ahead but has accepted the benefits you've described during the call, it is a good idea to ask if he or she has any further questions or additional needs before asking for the business. You might say:

- Is there anything else I should know about in setting up this account?
- Any questions?
- Have we covered everything?

EXERCISE

Check the situation(s) below in which you would use the skill of closing.

- ☐ You've identified and supported several gaps in the customer's portfolio. The customer has accepted the benefits you've described and indicated that he has no additional needs or concerns.
- ☐ You've discussed the first need revealed by a customer, and the customer seems impressed with the benefits you've described. You don't know if she has additional needs.
- ☐ Toward the end of a call, the customer says, "Sounds like these wines would be a good addition to our wine list."
- ☐ When asked if there's anything else she'd like to know, a customer says, "No, but I'm not ready to make a decision yet."
- ☐ You've supported three needs. The customer is smiling and nodding.

HOW TO CLOSE

There are six types of closes or methods of gaining commitment that occur at the end of the supporting stage or when you have overcome any concerns the customer may have shared. They are as follows:

1. *Summary-of-the-benefits close*, during which you review and summarize the key benefits the customer has previously accepted and ask for the order. This is a compelling and logical conclusion to make since the need benefits have been accepted and all concerns have been resolved by the customer. We will discuss this method more in the next section.

2. *Assumption close* assumes that the customer is willing to make the commitment. When a customer sends you strong buying signals, the salesperson using this method takes on the tone that the customer has given his or her verbal commitment by saying something like, "I can have your first delivery to you Monday. All I need is your signature on my order sheet." A response by the customer to the delivery date or willingness to sign the order sheet is a statement of commitment.

3. *Special-concession close* offers the customer something extra for his or her immediate commitment. This method should be used with caution since some customers may become skeptical as to its intent. Use it only where the customer is on the edge of making a decision and could benefit from a slight nudge. You might say something like, "I would like to leave with your first order today. As a way of thanking you for your business, I am prepared to have our production department print your revised wine list to save you the time and expense of doing it yourself."

4. *Single-problem close* involves a single concern or objection that stands in the way of gaining the customer's commitment. Using this closing method, you might say, "It seems that you like all three wines we have discussed today, with the exception of the limited number of cases we can provide for the one cabernet. If my director of sales can increase your allotment by one additional case per month, can I tell him we have an agreement?"

5. *Limited-choice close* includes a narrow list of choice options as a way of helping the customer make a decision. The customer may have accepted the benefits you can offer but may have concerns about, for example, their price. Similar to a single-problem close, you describe various choice packages that can get the customer to the price point he or she desires and you can provide. In helping him or her consider the options, (1) explain the different choice packages one at a time so you can assess the customer's interest in each, (2) cease offering options when you sense the customer has been given an ample selection, (3) remove options in which the customer shows little interest, and (4) narrow the list to a select few. You might say, "It appears that you like these two options the best. Correct? The pluses and minuses of each are . . . Which one would you like?"

6. *Direct-appeal close* involves asking for the order in a straightforward manner. Using this method you might say, "I can get this product for you by the beginning of next week; all I will need is your signature on my purchase order form."

MORE ON SUMMARY-OF-THE-BENEFITS CLOSES

One of the most effective closing methods salespeople across all business sectors advocate is the summary-of-the-benefits close. In essence, all closing methods share much in common with this technique, so it is appropriate to discuss it in greater detail. To advance a sale to its next logical step or to ask for the sale, you do the following:

- Review or summarize the previously accepted benefits.
- Propose the next step for you and the customer.
- Check for acceptance.

Reviewing Previously Accepted Benefits

In the course of a sales call or several calls, you may support a number of needs with a variety of features and benefits. Ideally, you make mental or written notes of the benefits the customer accepts.

The first step in closing is to briefly review benefits that have been accepted by the customer. Reviewing previously accepted benefits reminds the customer of the good things he or she can look forward to if a purchase decision is made (or steps toward such a decision are taken), and it lets you convey your confidence in the wisdom of moving ahead.

For seasoned purchasing managers, the review of accepted benefits can be brief. For the new wine buyer, spending more time reviewing the benefits will be warranted. You can position your review of benefits with phrases like these:

- Let me recap some of the ways we can help you reach your wine sales goals . . .
- As we've discussed . . .
- Let's go over the highlights of what we've talked about so far . . .

EXERCISE

Check the appropriate statement(s) below:

1. In the first step of closing, you review or summarize these items:
   - ☐ All the benefits you've mentioned in your discussion(s)
   - ☐ The benefits that have been accepted by the customer
   - ☐ The features that make your wines or support services unique
   - ☐ The needs that are most important to the customer

2. During a call, a sales rep introduced the features and benefits of a wine distributor. The customer accepted these benefits. Check the statement(s) below in which the salesperson completes the first step of closing correctly.

☐ "Let's recap what we've talked about so far. Though you prefer limiting your wine list to twenty-five wines, our inventory of over five thousand wines will allow you to make changes when your customers' preferences shift quickly."

☐ "You also liked having access to our growing inventory of organically produced wines that will appeal to your health-conscious consumers."

☐ "In addition, our three-days-per-week delivery schedule will ensure that wine inventories will remain appropriately stocked."

☐ "Lastly, you like our pledge to limit the wine labels to fine-dining restaurants, which means that your customers will not see them offered by retailers for less money."

PROPOSING THE NEXT STEP
FOR YOU AND YOUR CUSTOMER

Once you've reviewed previously accepted benefits, you propose next steps for you and the customer. For example, in closing a new account involving a grocery store chain, you might say, "I can have your first order delivered by Monday. All I will need to do is fill out your vendor partnership and quality assurance/control forms. Can you e-mail them to me?" or "I understand that you will make your recommendations to your purchasing committee Monday of next week. I'd like to schedule a call with you for Friday to discuss our proposal that I will be sending you and address any questions you may have. What time Friday would be good for you?"

Specifying what you would like the customer to do next ensures that he or she is clear about the commitment you're asking him or her to make. Saying what you'll do next demonstrates your commitment to working with the customer. It says, in effect, "Here's what I'm willing to do to move us closer to satisfying your needs." And it helps to be enthusiastic at this point. As one customer told us, "A good salesperson, first of all, believes in his wines and shows it."

Be careful not to make commitments that are significantly more substantial than those you ask the customer to make. If the customer isn't willing to take some action, his or her interest may not be sufficient to balance the time and effort you invest over time. The commitments you and your customers make should reflect the mutuality that characterizes any good relationship.

When you are closing a sale:

You might ask the customer to:
- E-mail you his or her vendor partnership forms
- Make a recommendation to the buying committee

You might offer to:
- Complete the vendor partnership form
- Provide the buyer information to make a persuasive presentation

When the sales process is continuing:

You might ask the customer to:
- Review a proposal
- Arrange a visit to the winery
- Do some internal "selling"
- Schedule another call with you

You might offer to:
- Prepare a proposal
- Schedule a visit to the winery
- Prepare the customer to "sell" internally
- Call again to explore the customer's needs further

According to one sales director, "A salesperson is doomed to disappointment if he or she depends on the customer to take the initiative and sign the order."

Check for Acceptance

After proposing the next steps for you and the customer, you'll want to make sure the customer accepts the plan you've outlined. You might say:

- What do you think?
- Can you arrange that?
- How does that sound?
- Are you comfortable with this?

EXERCISE

In the closing statement below:

- Underline the part where the salesperson reviews previously accepted benefits.
- Put brackets [ ] around the part where the salesperson proposes next steps for herself.
- Put parentheses ( ) around the part where the salesperson proposes next steps for the customer.

- Strikethrough the part where the salesperson checks for acceptance.

As we've discussed, though you prefer limiting your wine list to no more than twenty-five wines, you see value in having access to our inventory of over five thousand wines, allowing you to make adjustments if your customers' preferences change quickly. You also like having access to our growing inventory of organically produced wines that will appeal to your health-conscious consumers. All I need is to complete your vendor partnership agreement and quality assurance and control forms and get a completed order from you. Can you e-mail me both by Friday noon?

PREPARING TO CLOSE

You prepare to close a sales call at the same time you prepare to open it. As you consider your objective for the call and the ground you must cover to achieve it, ask yourself:

- What next step(s) would it be appropriate to ask the customer to take (assuming the necessary ground is covered)?
- What next step(s) would I be willing to take?

It's also useful to prepare a "backup" close—that is, to identify lesser next-step commitments you might ask for and make if you don't accomplish what you'd planned or if the customer is reluctant to commit to the next step(s) you initially propose.

LOOKING FOR BUYING SIGNALS

The clues that a customer is receptive to buying are of two types. They are (1) verbal buying signal cues and (2) nonverbal cues.

Verbal Cues

- How soon can I get a completed vendor partnership agreement form from you?
- It's been a while since I sourced a good prosecco from the Italian Lake Garda district.
- We will need to have our quality assurance and control forms completed by you before we can place our first order; is that a problem?

Nonverbal Cues—If You Are in a Meeting

- Reexamining the price sheet
- Picking up a pencil and starting to figure
- Leaving the room to talk to someone else about the proposal
- Subtle indications of interest such as nodding one's head, rubbing and holding one's chin, or leaning forward in one's chair with a friendly facial expression

WHEN TO CLOSE

After overcoming an objection:

CUSTOMER: Your wine costs ten dollars more per case than the competition.

SALESPERSON: Yes, but when you consider it turns over quicker without discounting, the wine's slightly higher costs work out to your advantage.

CUSTOMER: Yes, I guess you are right.

SALESPERSON: Great. I can have an order delivered to you by Monday. All I need is your signature on my order form.

WHEN THE CUSTOMER STALLS

There may be times when a customer has accepted the benefits you've described but is reluctant to take the step you propose or postpones making a decision. The customer might say:

- Things are kind of crazy right now. I need to hold off on that for a while.
- Sounds good, but it's too early to make a decision.
- I'm just not sure. Let me get back to you in a couple of weeks.

When a customer seems reluctant to move ahead, question to find out why. *You* might say:

- Is there something else we should talk about?
- What needs to happen before you are ready to make a decision?
- Can you tell me what your hesitation is?

If the customer has a concern, you might be able to resolve it. If the customer is willing to move forward but at a slower pace, you can propose a lesser commitment than the one you initially asked for. Try to get the best commitment the customer is willing and able to make that day. If you can't get a customer to make a decision or commit to a next step, try to get a date by which a decision or commitment will be made.

Be sensitive to the fact that the customer's hesitation may be that he or she is too busy or preoccupied with more pressing decisions at that time. There may also be a fear of failure (e.g., a poor purchase decision) that may be behind such hesitation. But if you are reasonably confident that no additional needs or concerns exist, it is best to maintain your composure and schedule a time to follow up. So, at times, you must be patient and persevere in working through such inertia.

INCREASING YOUR CHANCES OF CLOSING

1. Focus on the prospect's dominant buying motive(s). These are the prospect's "hot buttons."
2. Negotiate the toughest points. If your product or service has a weak point, make sure that it surfaces before attempting to gain commitment. (This is the topic of chapter 12.)
3. Avoid surprises at the closing. Salespeople should ensure that potentially surprising information is not revealed to the prospect at the last step.
4. Display self-confidence at the closing stage.
5. Ask the prospect to purchase more than once. Research has shown that 50 percent of all salespeople asked for the purchase once, 20 percent asked twice, and the most productive salespeople asked three times or more.
6. Recognize clues that invite you to gain the customer's commitment—these clues are often referred to as buying signals.

WHEN YOU GET A DEFINITE "NO"

Sometimes you get a "no"—the customer has chosen another vendor or decided not to address his or her needs right now. Following are some ideas for what to do when a customer says no:

1. Thank the customer for taking the time to speak with you.
2. If appropriate, ask for feedback so you can learn from the setback: What factors contributed to the customer's decision? What did your wines or organization lack? What did you do—or not do—that affected the decision?
3. If you think there's a potential for future business and want to maintain a presence with the customer, ask permission to stay in touch. Send articles of interest or invite the customer to relevant events, but don't make purposeless calls.

And remember, sometimes the best decision you and the customer can make is the decision not to work together. Perhaps your wines or support services do not really match up with the customer's needs. Maybe what the customer needs does not represent good business for your organization. It's better to get a "no" and walk away from a particular piece of business than to continually get "maybes" and prolong a relationship that won't be mutually beneficial. As one VP of sales told us, "Good salespeople have the ability to walk away from business if it doesn't meet their organization's needs."

Even when you don't make a sale, your commitment to a mutually beneficial decision will work to your advantage in the long run. The customer will think of you when needs arise that you can address.

SHOULD YOU CALL ON A BUYER
FOR WHOM YOUR WINES ARE NOT A GOOD FIT?

If, in your precall preparation or during your first sales call, you find that your wines or support services are not going to be attractive to the buyer, should you even stop in? We contend that it's worthwhile to make the personal connection but also admit that the products you have right now are probably not a good fit for what he or she is trying to do. That is a valuable sales call. It shows that you are a professional, you understand the customer, and you are laying the groundwork for a future sales call when you do have something that will fit his or her criteria. And it never hurts to stop in from time to time to keep that relationship breathing—even if you don't have a wine to offer. You might offer to invite the potential customer to an event that your company is organizing that is devoted to a cause he or she supports or offers a chance to connect with people he or she might like to meet . . . just keeping the door open for a time when you have that rare cabernet franc from the Loire Valley. Right now you have to admit you don't have anything to interest the potential customer, but when you do, you can then stop in and make the pitch to someone who is no longer a stranger.

DISCUSSION QUESTIONS

1. Think back to when you recently closed a sale. What kind of a closing statement might you write for that call?
2. Should you review previously accepted benefits?
3. Should you propose next steps for you and the customer?
4. Should you check for acceptance?
5. Should you include new items that you didn't discuss in person?

THE LETTER OF AGREEMENT OR CONTRACTS

What can and cannot be included in contracts and vendor agreements varies by state and country. As discussed in chapter 2, each country (in the case of the United States, each state) is considered sovereign when it comes to the regulation of alcoholic beverages. For this reason, nationwide distribution of wine requires a familiarity with the laws of any country in which you do business. Hence, it is important that a winery or distributor intending to expand to a new state engage a competent beverage attorney who knows the particular interests and limitations of the firm and the beverage laws of each state and federal government. For wine sales reps, it is equally important that they consult with their director of sales to determine what closing document is needed when adding a new account.

In the United States, each state's wine distribution law will fall into one of three basic categories: franchise states, nonfranchise states, and control states. Currently, twenty-two states allow wineries to create exclusive franchise agreements with one or more distributors in each state. Franchise agreements hold promise for creating purchasing partnerships allowing both parties to work collaboratively for mutually satisfying gains. In simple terms, a franchise state creates a long-term and binding relationship between the distributor and your brand. However, it should never be entered into lightly. Once an agreement with a distributor is signed by a winery, the distributor essentially owns the rights to distribute its wines in perpetuity. Terminating a poorly performing distributor may lead to costly litigation or penalties imposed by the state's alcoholic beverage compliance office. Nonfranchise states do not permit such exclusivity, allowing wineries to work with multiple distributors in the form of nonbinding buyer-vendor agreements.

Among the so-called control states, the state government controls the distribution of wines within its borders, acting as the state's wholesaler-distributor, selling wine at its state-owned stores. In this setting, the private for-profit distributorship model is replaced with the public sector, "even playing field," regulatory hand of local state government. In such states, there is no freedom to include in contracts any attempts to penalize the state for poor sales performance.

The relationship between distributors and retailers is confined to nonfranchise and control states. Retailers in nonfranchise states can and do enter into exclusive relationships with distributors, but these relationships, to our knowledge, can be terminated more easily than franchise agreements by either party. In addition, exclusive relationships are generally entered into by a retailer based on some incentive. For example, a grocery store chain, as an incentive to entering into an exclusive relationship with a distributor, will often receive full-time merchandising support, making the display and stocking of wines totally hands-free on their part. In addition, a restaurant owner may enter into an exclusive relationship with a distributor in return for free glassware. In nonfranchise states, most retailers will work with a number of vendors, managing their inventory themselves. In such settings, the distributor may require a credit check on the business prior to taking on the new account. Returning to the example of the grocery chain, the buying organization may require the distributor to complete a partnership agreement that includes quality assurance and quality control forms needed in the event that a recall of the wines becomes necessary.

For these reasons, planning and foresight can go a long way in avoiding problems down the road in franchise states, nonfranchise states, and control states. Though it is well beyond the scope of this book to give a full accounting of the appropriate regulations in all fifty states since they are continually evolving, it is essential that sales reps consult with their director of sales or state alcohol board on what is permissible by law in buyer-supplier relationships. Drawing up letters of agreement or contracts must fall within each state's legal statutes.

As mentioned earlier, when a salesperson makes a sales presentation that contains *supporting statements*, the salesperson is simply building the *proposal*. When presented as a written document (written RFP, eBid, or another document), it should look very similar to the final *letter of agreement*. In fact, if the buyer signs the proposal as written and/or with minor initialed changes, the *proposal* is formally transformed into a basic *letter of agreement*. Again, where permissible, a signed *letter of agreement* becomes a formal *contract*.

Elements of a letter of agreement or contract are highly proprietary. However, such buyer-vendor relationship agreements generally contain the following elements:

1. Termination/cancellation: What happens when either party wishes to cancel such an agreement? Remember that in franchise states, such clauses must conform to the state's alcoholic beverage compliance board's statutes governing franchise terminations.
2. Penalty clauses: What happens when either party does not fully live up to the terms of the letter of agreement?
3. Legal jurisdiction: It is preferable, and more advantageous, for a party of legal litigation to be subject to the legal jurisdiction where the party is located.
4. Arbitration clauses: Arbitration is a means for settling disputes outside the court system and is less costly. It is usually in the best interests of both the vendor and customer to use arbitration instead of the court system.

Chapter Twelve

Negotiating Customer Concerns

At any point in the sales process, a customer may raise an objection, voice a concern, or express a reluctance to make the commitment to your offer. It is important to recognize that if customers raise such a concern or objection, they are genuinely showing interest in what you have offered. If they had no interest, they would simply thank you for your time and end the conversation.

Therefore, you should welcome concerns and objections being raised—not fear them—as it's just one of the ways customers express their needs. Encouraging the free expression of concerns and facilitating an open exchange about them demonstrates your commitment to helping the customer make an informed, mutually beneficial decision. As one sales manager told us, *"A customer who raises a concern or objection is sending me a good sign. It means that the customer is interested and tracking with me. Objections give me an opportunity to talk further and gather and provide more information."*

This chapter introduces you to the types of concerns typically encountered on sales calls in the wine trade and how to resolve every kind of concern. We have reviewed a lot of training manuals, mostly focused on the retail level, that advocate overcoming customer objections by applying the correct manipulative technique aimed at outsmarting, outtalking, and outmaneuvering the customer. Unfortunately, they can be quite effective in retail environments where the market is first-time customers.

However, manipulative tactics seldom work in the business-to-business wine trade sales environment for two primary reasons. First, you are dealing with professional buyers who are not prone to purchase on impulse. They will do their due diligence in weighing their options among multiple suppliers since often so much is riding on making wise investment decisions. Second, in most buyer-supplier environments, your focus is not only gaining the first sale with a customer but also retaining your customer's loyalty in terms of

Figure 12.1.   Reassuring the Customer

Offering proof

repeat purchases and willingness to recommend. The high cost of prospecting new customers when compared to the value of the first sale often contributes little net gain for the selling organization. Profits are generated when customers reorder or rebuy as well as when they recommend your services to others.

As we have emphasized throughout this book, the consultative selling strategy is focused on building strong relationships with customers. To accomplish this, the salesperson must pursue a win-win strategy for both the buyer and seller organizations.

Consider for a moment that there are essentially three possible outcomes of a purchase decision for both the selling and buying organization:

- The seller loses, the buyer wins—game over.
- The seller wins, the buyer loses—game over (or I play next time to get even).
- Both the seller and the buyer win—relationship continues.

Win-win relationships are the foundation of healthy businesses and sound economies. They allow both the wine seller and customer to achieve their

Table 12.1. Three Basic Types of Customer Objection

When a Customer:	*You're Dealing with:*
Doubts a feature or benefit you've described	Skepticism
Thinks you cannot provide a feature or benefit you can provide	False impression
Is dissatisfied with the presence or absence of a feature or benefit	Shortcoming

goals in ways that enhance the relationship for each party. The approach is more than pursuing a simple compromise where concerns arise. Compromise does not always ensure that the outcome for both parties will be satisfactory. Producing win-win outcomes requires finding solutions to problems that are satisfactory to both parties. These solutions can be found by gaining a mutual understanding of the elements of the concern.

Remember, your customers do not need your wine. They have lots of other choices. But they will buy your wine if they are convinced it will continue to make them money. And if you want to be successful, you have to work through their concerns and reluctance to buy, and once sold, make sure that you deliver on your promises.

Generally speaking, customers express three types of concerns that make them reluctant to buy. They are (1) skepticism, (2) false impressions, and (3) shortcomings (table 12.1).

SKEPTICISM

When you make a supporting statement linking a specific feature to a customer benefit, sometimes customers raise doubts that your product or service has the features or will provide the benefits you have highlighted. And you need to pay attention to both verbal and nonverbal communication. One sales manager told us, *"The customer sends out signals. You notice them from his questions and comments. It's important to be very sensitive to these signals."*

The customer may say:

- I've yet to see a small wine distributor come close to that of a large national distributor regarding merchandising support and price.
- 24/7 merchandising support? Sounds too good to be true.
- Your chardonnay is a bit too heavy and oaky. I just don't think it will pair well with our food.

When a customer expresses doubt that your product or service can actually deliver what you have said it will, you've encountered skepticism.

FALSE IMPRESSIONS

Some concerns arise because a customer has incomplete or incorrect information about your product or service. Behind the concern is a need you can satisfy, but the customer doesn't know you can satisfy it—usually because the need has not yet surfaced and been discussed, hence you have not yet supported it.

For example, a buyer for a restaurant chain who doesn't know that your small wine distributorship carries a sufficient inventory might say, "I don't think your company has the capacity that will work for my group. We have always used one of the largest distributors in each state we operate in. Consolidating our purchasing power with one distributor who can handle all our in-state locations allows us to command the best price." When a customer has a concern because he or she thinks you can't provide a particular feature or benefit when, in fact, you can, you're dealing with a false impression.

SHORTCOMINGS

Any recommendation you make to a customer reflects your best efforts to address that customer's needs with your organization's products and support services. But every product and service has shortcomings or limitations, and you can't always satisfy all of a customer's specific needs. One customer told us, *"An effective salesperson is honest about the product. He admits what his organization can do as well as what it cannot do."*

When a customer has an accurate understanding of your products or services but is dissatisfied with the presence or absence of a feature or benefit, you're dealing with a shortcoming—a need you can't satisfy. For example, if a customer wants to do business with a wine distributor who can offer him or her a full-time merchandiser, the absence of this feature would be perceived as a shortcoming or a drawback.

When you encounter a shortcoming, it's important not to dismiss it as unimportant. Responding openly to a shortcoming demonstrates your integrity and reflects well on both you and your organization.

SUMMARY EXERCISE

Match each customer statement below with the appropriate concern:

- SK (skepticism)
- F (false impressions)
- SH (shortcomings)

1. I don't think your company has the capacity that will work for my restaurant group. We have always used one of the largest distributors in each state we operate in. Consolidating our purchasing power with one distributor who can handle all our in-state locations allows us to command the best price. (The buyer does not understand that you have a sufficient inventory spread across four warehouses in your state and are prepared to match the price of your largest competitors.)
2. It's hard for me to imagine that you can live up to your commitment of 24/7 merchandising support. Sounds too good to be true. (You can.)
3. You know, we don't see potential to take on a winery that produces fewer than twenty thousand cases annually. (The customer does not know the winery produces thirty thousand cases per year.)
4. How do I know your distributor can deliver the luxury wines my restaurant customers expect? (Your small distributorship carries only a limited selection of wines at the luxury level.)
5. Your Napa Valley cab is too expensive for us—in fact, it's almost 20 percent higher than cabs I can get from Chile. (Your price is 20 percent higher than a comparable wine from Chile.)
6. We're not interested in using wine distributors who cannot provide us with a full-time merchandiser at each of our stores. (You cannot afford this.)

USING QUESTIONS TO PROBE FOR AN UNDERSTANDING OF A CONCERN

When a customer first expresses a concern, the nature of the concern may not be clear. When you're not 100 percent sure what kind of concern you're dealing with, probe with questions until you are sure. Even when you do know which type of concern you're encountering, it may be important to explore with questions for a fuller understanding before responding:

- With skepticism, it will be especially helpful to understand what specific aspects of your offer cause doubts.
- With false impressions, as with any need you can satisfy, it's helpful to understand specifically what the customer wants—and why.
- With shortcomings, it's helpful to understand the need you can't satisfy— and the need behind it, if one exists. Even though you can't provide the specific feature or benefit the customer wants, understanding why that feature or benefit is important helps you plan and position your response.

Besides enabling you to respond appropriately and effectively, the use of probing questions to understand a concern allows customers to fully express their misgivings about your product or organization. The information they provide will be useful to you in responding to their concerns in an appropriate manner.

EXERCISE

A customer says, "I read your proposal, and I'm not sure that your group can handle an account of this size."
 What's the first thing you would do? What might you say?

Resolving Customer Concerns

Skepticism and false impressions are similar in one respect: in both cases, the customer has a need that can be satisfied with your product or service:

• With skepticism, the customer doubts you can provide the feature or benefit you have described in a supporting statement, but the feature and benefit do satisfy a need expressed by the customer.
• With a false impression, the customer thinks you can't satisfy a particular need that you can satisfy—often because the need has not yet been expressed, discussed, and supported.

Resolving Skepticism

A customer who expresses skepticism needs reassurance that your product or service has the features and/or will provide the benefits you've described. To provide that reassurance, you (1) acknowledge the concern, (2) offer relevant proof devices, and (3) check for acceptance.

1. Acknowledge the concern: In responding to any concern, it's useful to let the customer know that you understand and respect it. In the case of skepticism, you might say something like, "I appreciate that you would want to be certain about something as important as this." Or "Given your previous experiences, I can understand why you would require this."
 When you acknowledge customers' skepticism (or any concern), be careful not to suggest that you believe there are problems with your product or service when there are none. In your haste to show respect for their concern, you do not want to inadvertently say things like "You're right"

or "Many of our customers have that same concern." You simply want to convey that you can appreciate the customer's point of view.

2. Offer relevant proof: Similar to your use of proof devices in your supporting statement, you provide evidence that your product or organization does have the feature and/or does provide the benefit you've described. For example, suppose you told a customer that your company can easily serve all of his or her stores in your state. You might say, "Here's a map showing the locations of our four distribution centers in our state. As you can see, none of your stores will be farther than one hundred miles from our nearest distribution center, allowing us to fulfill all orders within twenty-four hours."

   In offering proof, make sure the proof you offer is relevant. When you have several proof sources, use the one that best addresses the specific feature or benefit the customer is skeptical about.

3. Check for acceptance: After offering proof, check to make sure the customer accepts it. If the customer rejects your proof, probe to find out why and, if possible, offer a different source of proof. Or ask the customer what kind of evidence could be acceptable.

EXERCISE

1. Your winery makes a big, heavy, oaky chardonnay, and the distributor who services fine-dining restaurants expresses a concern that it will not sell in his market or to his customers.

   Respond to this customer's skepticism by (1) acknowledging the concern, (2) offering relevant proof, and (3) checking for acceptance.

2. You are selling to a purchasing manager of a large grocery store chain in your state. When you make a support statement describing the variety of wines in your inventory, the customer responds by saying, "I don't believe your company has the inventory to support our stores." You have a current inventory list with you. You also feel that seeing is believing and are prepared to extend an offer to the customer for a site inspection.

   Respond to this customer's skepticism by (1) acknowledging the concern, (2) offering relevant proof, and (3) checking for acceptance.

3. Whatever you sell, it's important to know which features and benefits customers tend to be skeptical about and to identify credible proof sources for them.

Produce a list of proof sources that you believe would be particularly helpful to have with you when making a sales call with a customer at his or her office. We will start you off with a few, but add creatively to this list:

- Portfolio of wines in your inventory
- Fact sheets on wines you believe will be of most interest to your customers
- Samples of merchandising collateral
- Articles on your best wines

SUMMARY: RESOLVING CUSTOMER SKEPTICISM

A sales manager told us, "I assure my customers that my purpose is to educate them and help them get what they want." Here's an outline of how to do that:

First

- Probe to understand the concern.

When

- It is clear that the customer is skeptical about a feature or benefit you have offered.

How

- Acknowledge the concern.
- Offer relevant proof.
- Check for acceptance.

Resolving a Customer's False Impressions

A customer whose concern arises from a misunderstanding thinks you can't satisfy a need you actually can satisfy. To clear up the misunderstanding, you should (1) confirm the need behind the concern and (2) support the need:

- Acknowledge the need.
- Describe relevant features and benefits.
- Check for acceptance.

Confirm the Need behind the Concern

The first step in addressing a false impression is to turn it around—that is, to get the need behind the concern expressed as a need (something the customer desires) rather than as a problem (a limitation of your product or service). To confirm the need, use an open probe question that includes the language of needs. For example:

> CUSTOMER: To be quite frank, we don't see potential to take on a winery that produces fewer than twenty thousand cases annually.
>
> SALESPERSON: Why is that? (probe to understand the concern)
>
> CUSTOMER: Well, our experience has shown us that for the volume of wines we demand at the prices we need, small wineries cannot deliver on a consistent year-to-year basis.
>
> SALESPERSON: So you're looking for a reliable supplier who can deliver you the quantity you expect at a price you can resell profitably. Is that right? (confirming probe to transform the expression of a problem into an expression of a need)
>
> CUSTOMER: Exactly.

Once you've confirmed the need behind the concern, as above, you may need to probe further about the "what" and "why" of the need to be sure you have a clear understanding before you support.

EXERCISE

You are selling to a wine distributor who states, "We do not see potential to take on a winery that produces fewer than twenty thousand cases annually."

The customer does not know your winery produces thirty thousand cases per year and is in the process of expanding its production by 25 percent. Though your company has a strong wine buyers' club, the owner knows that growth will come by partnering with a few large distributorships that can sell its wines at higher volumes, albeit at a lower price.

The customer says, "I like the taste of your wines, but I need to go with wineries that have greater capacities and economies of scale. I just see no reason to drop a line of wines to carry another unless I feel that they can sustain themselves."

Respond to this customer's misunderstanding by confirming the need behind this concern.

Support the Need

Once you've confirmed the need behind a false impression and have a clear understanding of the need, you proceed as you would with any need you can satisfy—that is, you support it by (1) acknowledging the need, (2) describing relevant features and benefits, and (3) checking for acceptance.

EXERCISE

Suppose you confirmed the need of the customer in the previous exercise like this:

> SALESPERSON: So it is important that the winery has the capacity to keep its quality and production capacity high and prices low.
>
> CUSTOMER: Absolutely.

Respond by supporting the need by demonstrating that your owners are making a substantial capital investment in growing their capacity to serve the wholesale market.

SUMMARY: RESOLVING A FALSE IMPRESSION

As a sales manager told us, "It is important in terms of the customer's ego that the misunderstanding is not laid at the customer's door."

First

- Use probing questions to understand the concern.

When

- It's clear that the customer thinks you can't provide a feature or benefit that you can indeed provide.

How

- Confirm the need behind the concern.
- Support the need:
  - Acknowledge the need.
  - Describe relevant features and benefits.
  - Check for acceptance.

SHORTCOMINGS: NEEDS YOU CANNOT SATISFY

Shortcomings arise because a customer has a need you cannot address or satisfy. The customer is dissatisfied because your product or service does not have the desired feature, cannot provide the desired benefit, or has an undesirable feature.

When you encounter a shortcoming, it's particularly important to probe to understand the need(s) behind it—what the customer wants and why. Even though you can't provide the particular feature or benefit the customer is looking for, you can still position your response to show that the features and benefits you can provide will contribute to the overall results he or she is looking for.

Resolving a Shortcoming

A customer whose concern is based on dissatisfaction with what you offer (or do not offer) must weigh the importance of the needs you can satisfy against the importance of the need(s) you cannot. To help the customer make this assessment, you should do the following:

- Acknowledge the concern.
- Refocus on the big picture.
- Outweigh the shortcoming with previously accepted benefits.
- Check for acceptance.

Refocusing and Outweighing

1. Acknowledge the concern: As with the other types of concerns you've considered, it's a good idea to let the customer know that you understand and appreciate it. You might say, "I can see why this is important to you" or "Price is an important factor in all buying decisions these days."
2. Refocus on the bigger picture: Once you've acknowledged the customer's concern about a shortcoming, you want to help the customer put the drawback in perspective—to consider it within the broader context of his or her other needs. You can do this by inviting the customer to step back and look at the bigger picture. You might say, "I understand that controlling costs is important. These days, keeping a sharp eye on the bottom line is important. I understand that our wines are 10 percent higher per case than you can get elsewhere. Let's step back for a moment and consider everything together."

   In addressing a shortcoming, avoid using the word *but* between acknowledging and refocusing. If you say *but*, the customer may feel that you're minimizing the importance of his or her concern. Instead, use words and phrases like *and it's also true*, *let's look at*, or *let's consider*. Or ask the customer's permission to step back and consider the bigger picture.

3. Outweigh the shortcomings with previously accepted benefits: Once you've refocused on the bigger picture, you can sometimes offset a shortcoming by reviewing the benefits the customer has already accepted. This helps the customer weigh the essential needs that will be satisfied by your product or organization against the need(s) that won't. As one sales manager told us, *"I try to find out how big of an objection it is and put it in the context of product advantages."*

For example:

- We've discussed the potential advantages of offering your customers the option of purchasing organically produced wines that are bottled without the use of nitrates.
- You and your colleagues also like the taste of our wines and that they are produced locally.

When selecting benefits to review, ask yourself which ones would be most likely to outweigh the drawback. Consider reviewing the following:

- Benefits that address the highest-priority needs of the customer
- Benefits that address the need behind the need in some other way (e.g., wines that may appeal to a whole new segment of wine buyers who prefer organic wines)
- Benefits you know your competitors can't provide

4. Check for acceptance: After responding to a shortcoming, check for acceptance. You might say, "Given the advantages of offering an organic wine produced locally, don't you think you can pass on the slightly higher costs to your customers?"

EXERCISE

In the example below:

- Underline the part where the salesperson acknowledges the customer's concern.
- Put brackets [ ] around the part where the salesperson refocuses on the bigger picture.
- Put parentheses ( ) around the part where the salesperson tries to outweigh the drawback with previously accepted benefits.
- Circle the part where the salesperson checks for acceptance.

I appreciate that you require your wine vendors to provide you with a full-time merchandiser at each of your stores. Labor costs are important to control, and keeping your wine merchandising as hands-free as possible allows you to keep your costs low in order to meet your revenue goals. With that in mind, could we review some of the other aspects of our vendor service we've discussed that could have a favorable impact on your revenue goals? For example, we talked about our ability to work with your staff to pair and serve wines along with their scheduled food tastings. We agreed that this would be a unique personalized offer in your market that should increase not only your wine but also your food sales. In addition, you like our ability to provide you with point-of-sales tracking data that allows you to select only those wines with favorable inventory turnover rates. Given the potential of this offer to engage your customers on a personal level, and ability to pick only those wines that sell well, could you afford the modest efforts to keep two shelves stocked by your staff?

When You Cannot Outweigh a Shortcoming with Previously Accepted Benefits

The benefits a customer has accepted do not always outweigh the shortcoming the customer is concerned about. And sometimes a customer points out a shortcoming of your products or support services before you've had a chance to introduce enough benefits to outweigh it. When you cannot outweigh a shortcoming for either of these reasons, you will want to employ probing questions to uncover additional needs you can support with additional benefits. If the customer accepts enough new benefits, you can try to outweigh the drawback again.

Of course, a customer who is focused on what you cannot do may not be interested in exchanging more information with you. To overcome any indifference the customer might be feeling, proceed as you do when a customer is uninterested at the beginning of a call:

1. Acknowledge the customer's point of view.
2. Request permission to probe.

For example:

CUSTOMER: You seem to have a nice selection of wines but lack the merchandising services we require of our vendors. Anything less than a full-time merchandiser will not do.

SALESPERSON: I appreciate the importance of full-time merchandising support. No one will be happy if your wine shelves are not attractive and in good order. Could we move on to some other areas that are also important to you, with the understanding that we'll come back to address this later?

If you gain the customer's permission to probe, go on to probe to create customer awareness of other needs.

SUMMARY: RESOLVING A SHORTCOMING

First

• Ask probing questions to understand the concern.

When

• It's clear that the customer is dissatisfied with the presence or absence of a feature or benefit.

How

• Acknowledge the concern.
• Refocus on the bigger picture.
• Outweigh with previously accepted benefits.
• Check for acceptance.

EXERCISE

You are meeting with a wine specialty shop owner who twenty minutes into your meeting indicates she is interested in adding a chardonnay that scores ninety points or higher at the cost of no more than $120 a case. At that price, you can only offer her an eighty-four-point chardonnay at $110 per case or another ninety-point chardonnay at $140 per case.

Assume what you like in previous statement. Respond to this shortcoming by communicating the following:

• Acknowledging the concern
• Refocusing on the big picture
• Outweighing the shortcoming with previously accepted benefits
• Checking for acceptance

Yes, it is easier to sell based on the fact that your wine at the price point desired is better by at least some standard of measurement than the wine the potential customer currently offers. What do we recommend when we are selling a wine that is measurably not as good as what the customer has on the shelves? Do we offer a more expensive option, attempt to convince him or her of the quality of our eighty-four-point wine, or do we have to admit failure? Consider for a moment another door. Focus on wine style, focus on packaging, focus on marketing support, and focus on what the consumer wants in your attempt to either sell the eighty-four-point wine or a more expensive option. Again, it has to come back to whether or not the wine will sell through to the end customer and if that process will help make the buyer more money and earn more customers.

But there is a danger here. If we suggest that our eighty-four-point wine will outsell the wines on his or her shelf, the buyer may not take that lying down. The response may suggest that he or she is not only not interested in our wines but also that we are no longer welcome in his or her store. "The whole reason people walk into my shop is because they like what I have to offer—and those offerings are based on my keenly honed palate. Now you come in here and tell me that you've got a crappy bulk-produced wine that is for sale in every damn convenience store in the neighborhood and that somehow I need to carry it."

REVIEW EXERCISE: CUSTOMER CONCERNS

Match the statements below with the appropriate concern:

- SK (skepticism)
- F (false impressions)
- SH (shortcomings)

1. This type of concern arises when a customer doesn't believe that your wine or organization will do what you've said it would.
2. With this type of concern, the customer is looking for reassurance.
3. Respond to this concern by acknowledging it, refocusing on the bigger picture, outweighing with previously accepted benefits, and checking for acceptance.
4. It's essential to turn around this type of concern so that it's stated as a need rather than as a problem.
5. Respond to this concern by acknowledging it, offering relevant proof, and checking for acceptance.
6. This type of concern arises when a customer thinks you can't provide a feature or benefit you can provide.

7. With this type of concern, you want the customer to decide whether or not the needs you can satisfy are more important than the need(s) you can't satisfy.
8. Respond to this type of concern by confirming, then supporting, the need behind the concern.
9. This type of concern arises when your product or organization doesn't have a desired feature, can't provide a desired benefit, or has an undesirable feature.

PRECALL PREPARATION TO RESPOND APPROPRIATELY TO CUSTOMER CONCERNS

A salesperson should expect customers to raise concerns. Being prepared is the key. The following outline provides ways to prepare yourself to respond to concerns a customer might raise on a call:

1. Find out—by considering your experiences and asking colleagues and your director of sales—what customers tend to express doubts about. Then identify relevant proof sources you could use to reassure them.
   Research the competitors a customer may be talking to. Ask yourself:
   • What kinds of information do your competitors communicate to the customer that would lead to false impressions?
   • What features and benefits do the competitors offer that might lead the customer to express the shortcomings of your product or organization?
2. Remind yourself never to ignore or dismiss a customer's concern.

Chapter Thirteen

Selling to a Lack of Interest

Ultimately your job as a wine distribution sales professional is to sell in markets where supply often exceeds demand. Hence, it should come as no surprise when you meet customers who have no interest in purchasing the wines you sell. Given the consultative needs-based approach to selling that this book advocates, without first identifying a need the customer is willing to share, there is nothing the sales rep can do to support. Therefore, the appropriate next step when faced with "no interest" from customers is to thank them for their time and ask if you may periodically contact them when new opportunities arise.

However, we have witnessed great sales reps who are prepared to tactfully respond to their customers' lack of interest in ways that ultimately serve the interests of both buyers and sellers. Though the method offers no guarantee for success, it is effective and should be added to a sales professional's quiver of approaches.

Selling to a customer's lack of interest builds on the core consultative selling techniques discussed in the previous chapters. However, we often see that what separates top sales performers from the second tier is that their customers see their interaction with the sales rep as a source of value separate from his or her products and support services. In essence, these sales reps are taking the initiative to teach their customers something new and different in ways that add value to the customers' business. Sometimes the sales rep will attempt to inspire new ways of thinking, and sometimes he or she will challenge customers' status quo with the goal of redefining their reality and potential for their business.

A customer may show a lack of interest, believing that adding new wine labels will not provide an increase in wine sales but simply pirate away sales from existing labels on the shelves (i.e., sales shift instead of sales lift). The customer may also simply be satisfied with his or her current wine labels, assuming that future success will follow from maintaining consistency with the status quo. Remember, telling a customer he or she is wrong or needs to see things differently will rarely produce satisfactory outcomes. "A man convinced against his will is of the same opinion still" is an old saying worth remembering. With that said, wine buyers we have interviewed indicate that a good sales presentation filled with useful information, when applied honestly and tactfully by a sales manager, is the type of advice and counsel buyers value from their suppliers. They lead their customers toward new ways of thinking with ideas and solutions that matter.

PROBING FOR OPPORTUNITIES, NOT NEEDS, THAT YOU CAN SUPPORT

This approach builds on the consultative selling approach discussed in detail in previous chapters, with one basic difference. In a call opening where a customer indicates that he or she is not interested at this time in considering adding new wine labels to the inventory, the sales rep simply acknowledges this fact but then follows up with a probing question to identify opportunities that will evoke a need that the customer had previously not considered.

Consider the following illustration:

CUSTOMER: Thanks for your call, but I am happy with my wine selection at this time.

SALESPERSON: I totally understand. You know what works for your restaurants. However, I did want to make you aware of a special on a chardonnay I have right now. It scored ninety and I have it for seventy dollars a case. Would lifting the quality of your house wines at a price that will allow you to recoup your costs with the first pour be worth a look?

Consider another illustration:

CUSTOMER: Thanks for your call, but I am happy with my wine portfolio at this time.

SALESPERSON: I totally understand. You know what works for your stores. However, I did notice that your selection of organic wines that you have on the shelf at the price point of nine dollars per bottle is limited. My company's tracking data have shown that organic wines have been leading growth in sales this year. Specifically, organically produced varietals at this price point have experienced year-over-year sales increases of 17 percent, without discounting, in markets very much like yours. I sincerely encourage you to take a look before our inventory is sold out.

In both these illustrations, the sales rep was not thrown off by the lack of interest but instead was prepared with a question designed to uncover an opportunity that can be supported. The sales manager's precall research may have provided insights as to what opportunity may lie with the customer who shows a lack of interest. A missing varietal at a price point on a shelf or poor wine-food pairings on a restaurant's wine list may yield opportunities to evoke customer interest that a sales rep can later support. We have also seen end-of-year rebate programs, complimentary glassware, and increased merchandising support used to induce a second look by a customer who initially demonstrates a lack of interest. However, selling to a lack of interest does not need to take such expensive forms to be effective.

PREPARING TO RESPOND TO A LACK OF INTEREST

As discussed in chapter 10 on supporting, sales managers should prepare by being able to translate features into benefits for the wines and support services they sell. For example, in preparing to respond to a customer's potential lack of interest, a feature-benefit worksheet should add potential probing questions designed to identify opportunities that can be supported (see table 13.1).

Selling to a lack of interest can also improve the chances of success when selling to a wine distributor, importer, or (in the case of Europe) large chain retailers. Given the potential volume and gross sales in winning such a new account, considerable effort should be devoted to researching the potential buyer's current portfolio, current suppliers, and any circumstances affecting his or her business relative to his or her wine inventory. Having a reputation of offering quality wines that are stable from year to year as well as providing

Table 13.1.   Presenting Features as Benefits

Feature	Benefit
A wine rated with a ninety-plus score at seventy dollars per case	A wine that your consumers will consider to be of good value
A wine organically grown and bottled at nine dollars	A wine that will appeal to the growing number of health-conscious consumers at a price point that sells
Three-days-a-week delivery schedule	With appropriate planning and forecasting, assurance that wine inventories will remain appropriately stocked
Wine exclusively offered to fine-dining restaurants	Your customers will not find the wine offered at retailers for less money

Table 13.2. Probing for an Opportunity That Can Be Supported

Benefit	Feature	Probes in Search of Opportunities to Support
A wine that your consumers will consider to be of better value	Wines rated ninety-plus points at seventy dollars per case	1. I noticed after reviewing your wine lists, your house wines are rated no higher than eighty points.
		2. Do you ever get pushback from your customers desiring a better-quality single glass?
		3. I have several wines in my book that are rated significantly higher at attractive prices.
A wine that will appeal to the growing number of health-conscious consumers who value local growers	A wine organically grown and bottled locally	1. Have you experienced increased demand for organically grown and locally sourced meats and vegetables?
		2. My clients who have responded to this trend have also had great success in adding locally produced wines to their wine lists as well.
		3. I have several wines in my book that are locally grown organically at prices I believe you will find attractive. May I show you one?
With appropriate planning and forecasting, assurance that wine inventories will remain appropriately stocked	Three-days-a-week delivery schedule	1. How much space can you dedicate to your wine inventory?
		2. Do you ever run short of certain labels and varietals?
		3. Does this lead to lost customers?
Your customers will not find the wine offered at retailers or casual restaurants for less money.	Wines exclusively offered to fine-dining restaurants	1. Do you know where else in this town customers can purchase wines that are on your wine list?
		2. Has the nonexclusivity of your wine list ever caused problems for you?

reliable support services can attract such potential buyers to reach out to you. However, even with a strong reputation, you are seldom without competition. Doing your homework in assessing your customer's needs prior to a call not only helps you gain that first meeting with an attention-getting opening but also allows you to distinguish your products and services from the competition. Consider the following illustration:

> CUSTOMER: Thanks for the call. I am sure your wines are great, but we are not interested in taking on any new labels this year.
>
> SALES MANAGER: I appreciate your directness and respect your decision. However, I did put in some time analyzing your portfolio and found some gaps that we can fill that I believe will add value to your range. I simply wanted to offer them to you first since I believe they will work well for you.

Large prospective buyers will have an expanded list of underlying needs and interests when compared to smaller accounts. Quality, packaging, and promotional support should be assumed to be a priority. Offering a wine that creates a new price point can be an added bonus. Wine labels that have achieved certain certificates, such as HACCP, ISO, and (in the case of imports from developing countries) fair trade certifications, can garner a second look from a reluctant buyer. The point is that selling effectively in a buyer's market requires preparation and tactful probing for opportunities that will induce interests in customers that you can support.

SUMMARY: SELLING TO A LACK OF INTEREST

You may not always find customers who wish to change their status quo. Too often one's clients are so overly embroiled in their day-to-day operations that they are not aware of recent trends in wine offerings or sales, nor are they aware of their competitive position. Wine reps who have a sincere interest in helping their clients succeed have an obligation to teach and occasionally challenge their clients in ways they can help them do better. The prepared sales rep will come ready to sell through an initial lack of interest by probing for opportunities that add value to the customer that the sales rep can support. A sales rep's time is money, so preparation for a sales call should be in proportion to the potential sales volume and revenue of winning the account.

DISCUSSION QUESTIONS

1. Which of your customers might be least interested in selling the most popular items in your portfolio? Why?

2. What can you do to overcome the resistance of a top sommelier at a key restaurant account?
3. What services or terms can you offer to help overcome sales resistance to your wines?

Section III

WINE TRADE

In the first two sections of this book, we addressed the role of sales in the wine business and the strategies and tactics of consultative selling in the world of wine. In this last section, we discuss the role of merchandising in wine and then go beyond the immediate issues of selling wine to explore how the techniques you have learned in the previous sections can be applied in the future. That application is valuable not only in your current position but also in your career as you advance within a company or beyond the limitations of your current employer. Consultative selling is a skill that reaps rewards both in the near term, by increasing sales, and also in the long term, by increasing your value as a key member of a wine sales team.

Chapter Fourteen

Merchandising

Every basic marketing text will start with the four elements that are always combined to create brand power in the marketplace: product, price, promotion, and placement. While the first two may not be within the realm of a salesperson to control, the latter two are critical elements to any major sales effort. Where you sell your wine and what you do to encourage sales in the market are two areas where you can differentiate yourself as an excellent salesperson.

Let's talk about placement first. In this chapter, we won't discuss the larger issue of selling wine exclusively via on-premise or off-premise or whether you should sell your wine in large discount stores or supermarkets. This was covered in chapter 4 in our discussion of sales channels.

In this chapter, we are going to specifically address the merchandising of your products in a retail shop or a restaurant. How important is this? Let's look at a couple of examples:

- If your Napa Valley chardonnay is mis-stocked on the shelf with all the inexpensive box wines, your sales will plummet. The customers who look for wine on those shelves are not interested in buying a much more expensive product, and it won't occur to the people who are looking for a top-quality chardonnay to seek it in that section.
- On the other hand, if you have just won a wine-of-the-year award from a top wine publication and you can convince a retail shop to create a large display on the end of an aisle during the holiday season, your sales will take off like a rocket. Those who are looking for your wine can't miss it in that highly visible location, and even those who weren't considering your wine may think again when they see it next to the rave reviews and suggestions that it's the perfect wine for a holiday gift.

161

- If you have a third hand, consider the difference between a crude floor stack of your most expensive wine compared to an elegant display of a few bottles accompanied by a sign that reads "Limited release—only two bottles per customer!" Which do you think is more effective at selling that wine?

Your job, as a salesperson, is to use the tools at your disposal to work with your retail and restaurant customers to make sure that your wine sells through as quickly and profitably as possible. With that in mind, managing its placement in the store and giving it the right kind of in-store support or promotion should be key elements in your work to service every account.

WHAT IS MERCHANDISING?

Merchandising is anything that goes beyond the wine bottle and label itself to help promote the wine in a retail sales environment. A neck-hanger or shelf-talker that gives the customer additional information and motivation to buy the product is an obvious example, as is a table tent in a restaurant that suggests a specific wine and food pairing as a starter course. A full list might include everything from store layout and product location in the store to product location on the shelf, in the cold box, displays, point-of-sale (POS) cross-promotions with other products or departments in the store, special pricing, and even regional or national advertising campaigns. For that matter, even the décor of the winery tasting room, or the parking lot, can be factors in how your customers respond to your product offerings.

To understand why these are important, we have to understand how wine consumers make their buying decisions. According to research by industry leader the Nielsen Company (2012):

> Wine drinkers are explorers and make their purchase decisions in-store. Compared to the beer and spirits categories, a high level of wine purchase decisions are made in-store (37 percent), and consumers make 70 percent of their product decisions at the shelf. Engagement with the category begins even before visiting the store. Wine samples, engaging in word-of-mouth and recalling exposure to advertising can greatly help boost this category's sales.

Given that the consumer makes 70 percent of product decisions at the shelf, merchandising is critical to your success. But wine retailers and restaurants have a limited amount of space, and every move to give one product more visibility almost certainly means that another product will be less visible. In fact, many stores track their sales per square foot, or per shelf facing, as a way of managing the products they carry. Which products are they going to promote? The ones that generate the most sales and the highest level of long-term satisfaction.

How important is the location of the product in the store? Look at any grocery store and you will see that the fresh products are always along the outside perimeter. There are two reasons for this. One is that the fresh products (meat, produce, and deli) require more attention throughout the day, and keeping them grouped along the perimeter makes this a lot easier for the store staff. But equally important is this simple logic: grocery stores put the most popular items far from the front door so that customers have to walk through the store and see all the other attractive products for sale. Keep those two ideas in mind as you think about merchandising your wine. You want the store staff to think your ideas make their life easier, and you want to make sure that customers see your wines for sale.

The most basic kind of merchandising is simple. You want your product to be visible on the shelf, right in the heart of the category where it belongs. You want the bottles to look clean and polished, not old and dusty. You want to have at least as many bottles on the shelf as your competition, and ideally, you'd like more of these facings, because they subtly communicate to the customer that your product is popular and deserves a bit more attention from both the store and the customer. Some larger wine companies have sales associates (i.e., merchandisers) who have only one job: to make sure that the company's wines look great on the shelf and that all the collateral merchandising materials are in place and working well.

Beyond the actual bottles on the shelf, any merchandising effort should make sure that your product stands out from the crowd. A little tag on each bottle that tells consumers that you recently got a great review is a basic type of merchandising. But now that almost every winery can and does produce these, they don't have the same impact. If every bottle has a tag, then none of the bottles stand out. Some stores embrace this and like the idea of every bottle getting a tag. Others have decided that the tags are nothing more than shelf clutter that has a negative impact on sales. They won't use them at all.

Good salespeople will include merchandising as part of their initial review of the account. They will notice how the store is laid out and which items seem to get more attention as impulse buys. Since many consumers buy wine on impulse, those displays by the cash register or next to the frozen pizzas may help sell a wine that wouldn't be as successful on the shelf right next to fifteen of its closest competitors.

The same is true in a restaurant setting. A wine that is promoted as the perfect pairing for fried calamari might get a lot more attention on a special page in the wine list rather than just being listed among the white wines for sale. A wine that is featured as a by-the-glass selection will often convince reluctant diners to experiment with something new and different—something they might never try if they had to order a full bottle off the wine list.

Conversations about these topics are a key part of providing good service to your accounts, especially if you can share a success story about another account that executed a successful promotion or merchandising campaign with your product. Ideas for increasing the sales of high-profit products are always welcomed by your accounts. And plans for decreasing the sales of low-profit products are almost as attractive. Sharing that type of information can not only lead to more sales but will also encourage your accounts to seek your advice and counsel on future merchandising decisions.

HOW DO YOU INCREASE THE SALES OF YOUR HIGH-PROFIT ITEMS?

You begin by giving them more shelf space or better placement on the shelf. Eye-level shelves are the most valuable spot in any store, while those shelves near the floor are the least valuable. In fact, many studies have shown that you will reduce sales of any product if you move it from eye level to the floor, sometimes as much as 45 to 50 percent. And moving items from the floor to eye level can increase sales up to 90 percent. Items that increase their shelf facings consistently increase their sales by as much as 80 percent.

A recent study by Gallo Winery noted that women who shop for groceries with small children in tow tend to avoid the wine aisle for fear that a little one might knock down a few bottles. For that target market, an end-aisle display might be the difference between sales and no sale at all.

Is there any doubt that you need to manage this in your accounts? Most retailers are well aware of this situation and will undoubtedly have some ideas on how they want their merchandise to appear. Before they add more spaces on the shelf for your product, they'll want to make sure that the new products add variety to the inventory. They'll want to make sure they are consistent in quality and continuously available. Most of all, they'll want to make sure that the products sell quickly, increase profits, and drive long-term loyalty to the store.

You can also place the product as part of a display or highlight it in the cold box. Top-selling products may also generate more sales when you extend the product line and add new items from the same producer onto the shelf. And then you can help promote these wines even further with POS displays and cross-merchandising campaigns. Sometimes, by placing your product next to an extremely popular competitor, you can encourage the customer to compare the two, and if your pricing and value proposition are effective, you can win that battle.

The easiest way to increase your white or sparkling wine shelf facings in any store is to make sure that you are not only on the shelf at eye level in the appropriate category but also that you are in the cold box along with the smaller group of wines featured there. Of course, this space is limited, as refrigeration is expensive. But it also appeals to those customers who want something to drink immediately, either at home or at a party. That's a good group to have as customers. And that's why you want to fight for a spot there—wine sales in the cold box are often three times as high as on the normal wine shelf. And in any cold box, wines are likely to be the most profitable item for sale. A wine that sells well in the cold box will make any retailer happy.

On the other hand, if you want to decrease the sales of less-profitable items, you should do the reverse: decrease shelf space, move them to a less-desirable location, reduce the number of products, and avoid wasting key space, time, and effort in promoting the products that aren't as profitable.

One final note about shelf management: the single best way to decrease sales of any product is to make sure it isn't in the store. If your product is not on the shelf or in stock at the restaurant, it won't sell a single bottle. That means every good salesperson needs to make sure that deliveries are timely and that customers get the wines they need at all cost. And it means that if you don't see your product on the shelf, but you know that it is in inventory, you should draw this to the attention of the manager. Telling a manager that your wine has sold off the shelf is not a bad message to convey because it draws attention to the fact that your wine is selling.

And if you want to do more to sell your wine? Work with your marketing team to develop effective sales materials that give the customer a reason to think about your wine. In the old days, this meant the usual shelf-talkers and table tents. But today, with smartphone-savvy consumers shopping in our grocery stores, it also means making sure that your website is fully up to speed and that your search engine optimization gets those consumers the information and messages you want them to have as effectively as possible.

TYPES OF POINT-OF-SALE
COLLATERAL FOR MERCHANDISERS

So what are the types of merchandising POS that you might create? We've adapted this explanation from the book *Wine Marketing and Sales* (Wagner et al. 2011) from the Wine Appreciation Guild:

- Wine factsheet: These are the bare facts about the wine, organized in a way that lets everyone in the distribution network get a quick picture of what makes each wine unique. These should include everything from vineyard and winemaking information to positioning, appropriate food pairings, and a reproducible label.

- Shelf-talkers: These are merely cards that appear on retailers' shelves with more information about the wine. The report usually includes a top rating but can also have tasting notes or wine and food suggestions. Some stores love these, while others aggressively prohibit them.
- Table tents: The on-premise version of the shelf-talker is the table tent, a simple card that is placed on restaurant tables to encourage customers to order your wine. Most restaurants will expect some special pricing allowance or volume discount in return for using your table tents.
- Neck-hangers: If you can't get your sales team to put the shelf-talkers up in retail stores, sometimes you can place the same information on a card that fits around the neck of the bottle. You can significantly increase the likelihood that these will make their way into the marketplace by putting them on the bottles at the winery before the wine is shipped. However, this involves increased labor costs as well as logistical challenges.
- Waiter cards: A simplified version of the wine factsheet, these are smaller and fit into a shirt pocket so that waiters can quickly and effectively represent the wines on the wine list.
- Coupons: Coupons are illegal for wine in many states, so be careful about how you use these. The classic theory behind coupons is simple: the coupons encourage consumer trial of the product. That trial then leads to brand loyalty. However, the major problem with this theory is that many people who use coupons only buy products offering coupons. You don't get brand loyalty; you attract coupon clippers. Since brand loyalty doesn't exist in the classic sense in the wine industry, coupons can't deliver it. They are now usually used by wineries to offer a temporary price reduction to gain market share, with the hope that the price reduction doesn't imply a lower price point for the wine.
- Case cards: Formed to fit behind a case to draw attention to your wine, a case card can send many messages, from simple statements of quality or style to complicated promotions involving coupons, tear-off recipe pads, or sweepstakes programs. As with all POS, each store will have its policy on these. Case cards are more complicated, expensive, and difficult to distribute than smaller elements of POS.
- Floor stacks and end-aisle displays: These are the holy grail of merchandising programs and reward retailers who buy in huge quantities. The displays themselves can range from complex installations with moving parts and life-size figures of celebrities to a simple, bold graphic on the shipping

case that creates a large image when the cases are stacked. These displays are tremendously effective, as they carry the implied endorsement of the retailer as well as some price reduction on the product.

The list could go on forever. It could also mention cold-box stickers, sales sheets, display bins or wine racks, video presentations, banners for the store windows or walls, cut-out standing figures, moving mechanized displays, QR codes, and even the design of your case shippers. When you are creating a stack of boxes, that design can serve as its display if it's done creatively. That means you can "create" a floor stack any time you need one.

Point-of-sale materials can play a considerable part in wine sales, and the most powerful of all are the floor stacks and end-aisle displays. But these are highly competitive and generally provide short-lived impacts. No store or restaurant is going to want to feature the same items all year long. That means you need not only good POS but also a successful pricing strategy that will get the attention of both the account and the consumer. And you need to present it with enough lead time so that you can execute it effectively. To be successful, you will not only need to sell the promotion but also the profit potential of the floor stack and even the ideal location in the store.

Where should you put a floor stack? The best locations are where there are a lot of customers but also where those customers are likely to slow down and think. The end of an aisle is obvious, as are the shelves directly in front of the cash registers. Other locations are in the meat or produce department, where you can sometimes gain added sales by suggesting an attractive food-and-wine pairing with the ingredients from that department. Any time you can display your wine in a stand-alone display away from the rest of the wines, you will see your sales increase.

How big should the floor stack be? A good rule of thumb is to estimate the total wine sales for the promotion and then stack that much wine in the display. While a larger display may look good on paper, if that display is still full of wine six weeks later, it will convince the store manager that it didn't work. That's not the message you want to convey.

In the end, how you present your merchandise will have a significant impact on how well it sells. And by using the elements presented here, you can not only give your products the help they need but also generate a competitive advantage in the marketplace over your competition. Of course your wines taste better. And now they will have the shelf presence and merchandising they deserve to get the attention of your account and the end customer as well.

DISCUSSION QUESTIONS

1. Which retailers in your area are most likely to be receptive to a large merchandising display? Why?
2. Go into a wine retail shop and look at the merchandising there. Which one do you think is most effective?
3. Now ask the store manager which one works best. Is there a difference? Why?

Chapter Fifteen

Strengthening the Relationship

As we have noted in previous chapters, our business-to-business customers have limited space in their portfolios, in their stores, and on their wine lists. In order to sell a product to them, you have to fight for space. In most cases, any new product is going to have to displace an existing one. There is a finite amount of shelf space in the store, and it's a zero-sum game. Your job as a salesperson is to convince your customers that selling your product will be more profitable than selling someone else's product.

But the goal of consultative selling goes far beyond that initial sale into the account. You eventually want to become a trusted supplier—a confidant or a consultant—to your accounts. To become a trusted seller to a retailer (or a distributor) you are facing the same kinds of challenges: a limited or finite amount of time on the part of customers. And so you have to convince them that spending time with you is more helpful and profitable than spending time with your competition.

We've talked about how to do that on the initial sales call in some of the previous chapters: do your homework, prepare well, be professional in every way, and make sure that you don't waste your or the customer's time. In this chapter, we expand on the concepts of consultative selling to help develop deeper and longer-term relationships with your customers.

BUILDING RELATIONSHIPS WITH YOUR CUSTOMERS

If this sounds a bit like dating, that's a good thing. Building relationships as a salesperson *is* a lot like dating. The first rule of dating someone you like is to make sure that you always get another date. One way you do that is to try to be as attractive as possible to the other person. Smart salespeople know that it's always a good idea to leave each sales call with a reason for a follow-up conversation about something of interest to the client.

INSERT GOOD PROBING QUESTIONS

One critical step in this process is learning to listen rather than sell. All too often, salespeople are too focused on pitching the product and not focused enough on what the customer is trying to say. If you have excellent listening skills and practice the art of active listening, you will find that you learn a lot more about your accounts. And that will help you understand what they want from you. By asking probing questions that really encourage a dialogue about the operations and future of the account's business, you'll be even further along the way. Listen. Ask questions. Take notes. Your goal is to go beyond being just another supplier and become a trusted friend to the account. Just like dating. (Except the notes. Don't take notes on a date!)

During your sales call, it makes perfect sense to answer most of the questions you are asked immediately and directly if you know the answer. But if you don't know the answer, that's a perfect chance for you to promise to get back to the buyer with the information he or she wants. And if you are unsure of the answers, it's still a good idea to communicate what you know and then promise to follow up with a confirmation to make sure.

In this case, you're doing more than just providing information. You are proving that you take the person's business seriously. Even better, you are laying the groundwork for an ongoing relationship built on you providing reliable information to your customers. That's the essence of consultative sales.

This isn't something that you should do only in response to questions from your clients. If you know that you are going to have information in the future that would be of interest, you should offer that during the sales call and then follow up with the knowledge that you mentioned during the call. For instance, you could provide the following types of information:

- Details about future products, pricing, and promotions that your company plans to announce in the near future
- If you know the account likes a specific vintage or lot of your product, let him or her know as inventories get low so that he or she can order the desired items before you run out
- Information about developments in the market as it becomes available to you—from the opening of a new restaurant to the closing of a competitor's store or the changing of a winemaker or wine style at a winery
- If you know one of your products will be featured in a magazine or newspaper article, send a copy of that article to the account the minute it appears

On a particular note, we would like to stress that no customer likes to read about significant developments at one of their suppliers via a story in the newspaper, online, or from a competitor. When you have major news about your own company, something that is worth a press release or a story in a magazine, you absolutely must get that information to your customers as quickly as possible and directly from you or someone in your management team. This includes everything from a change in label or winery name to the fact that you are no longer making a particular wine or that you have changed your distributor in the area. As the primary contact for the winery with your accounts, it's your job to keep them informed and to do so ahead of the rest of the world.

That is part of a larger role for any consultative salesperson. Of course, you have products to sell, but if you want to develop your relationships with your accounts beyond that, you need to offer things of value beyond the products themselves: industry news, marketing knowledge, thoughtful analysis, and memorable stories. We have already addressed industry news above, so let's talk about the other items on that list. If you work for a large wine company, you will have access to top-quality market research from companies like IRI and Nielsen. But most of your accounts won't have that available to them. By studying the reports, you are sure to find items, statistics, and trends that will help your clients. By sharing that information, you'll build a much stronger relationship with your accounts. (We should note that not all accounts find this information particularly valuable. The lone fine-dining restaurant in a small town might not share the same customer base or customer motivation that the other businesses in that town have—and the same is true in a larger market.)

How you do this may depend on what your customers prefer. Some would like to get a quick e-mail note from you about a trend that could affect their business. Others might prefer a phone call or ask you to hold the information for your next sales call. Remember, you are dating these people. They get to set some of the ground rules on how you will proceed. If one of your products does get an excellent rating or review, it helps to share that in some form that they can use in the store. Writing a text message to them about it may get the news out, but they would probably prefer to get a nicely laid-out sheet that they can hang right next to your bottles on the shelf.

What kind of analysis can you do? Nielsen has noted that Hispanics in the United States do not drink as much wine per capita as the rest of our population, but they do tend to drink more sparkling wine per capita than other Americans. That's something you might suggest to a wine shop owner or a restaurant that seems to have a lot of Hispanic customers. As marketers, the authors of this book are big fans of mining the data to get information that can give you a leg up on the competition.

The last item on our list above is providing your customers memorable stories about your wines. Far too often, wines are sold based on price, production methods, and ratings rather than memorable stories. That's ironic because very few wine consumers buy wines because they come from volcanic soils or have undergone malolactic fermentation. Instead, consumers fall in love with the story behind a wine: not just that the vineyard is farmed organically but that the owner's children often sleep out in the vineyard during the summer and watch birds and rabbits come out with the morning sun. While these sorts of stories work well in a sales call, they are also worth sending as a quick e-mail, Facebook post, or tweet to keep your winery and wines in the minds of your accounts. If you have listened carefully during your sales calls, most of your accounts will have let you know, either directly or indirectly, about some of the things they think are important. If they mention organic wines, sending them a photo of the owner's kids camping in the vineyard is a perfect follow-up.

As a salesperson, your primary goal is to sell wine to your accounts. But we suggest that by becoming a valued member of the industry for that customer, you can not only increase your sales but also enhance your professional relationship. Does the store host wine tastings? Then you should make an effort to attend some of them. You'll not only show your support of their efforts to educate consumers but you'll also learn more about how they work with their customers. If the tasting is a success, offer to host a similar event with your wines. Even if the event is not a smash, it gives you an opportunity to suggest an alternative (perhaps one that you know worked better at a different location) as a solution.

Do your customers hold staff trainings? Do they host winemakers in the store to sign bottles? Does the restaurant hold winemaker dinners for its top customers? All of these are opportunities for you to participate in their activities and hopefully begin to place your products in these events. If your winemaker is in town to meet with a critical journalist, that's an excellent chance for you to suggest just the right one of your on-premise accounts that will know the value of having that lunch in his or her restaurant.

As you begin to develop your relationships with your accounts and others in the wine industry, you can begin to see how this network will pay benefits of its own. If you have studied in one of the wine certification programs in the United States, you likely will know a few of the instructors, whether they be masters of wine, master sommeliers, wine educators, or even leading figures from other sectors of the world of wine. Keep track of these contacts, and take the time to drop them a note every once in a while to let them know what you are doing and how much you value their work.

There's a chance that when one of those industry leaders comes to your market, he or she will let you know. And that's the kind of information you can then use to good effect with your accounts. Will there be a small wine tasting hosted by the VIP? Can you bring a "friend" to that tasting? Is the VIP interested in finding a good restaurant for a quiet dinner? Can you suggest a few that might work? All these things are far beyond the scope of a basic salesperson, but they are absolutely part of the landscape of wine sales. Your competitors are doing exactly these sorts of things, and your job is to do them better.

This is not just limited to the wine business itself. If you develop the relationship to the point that you are sharing interests outside of wine with your accounts, you should absolutely look for ways to include these in your conversations. A quick text message about the baseball game last night would elicit a chuckle from a big baseball fan. Riding in a charity bike ride to benefit the American Diabetes Association might have a much more significant impact on an account that is promoting the ride. If a restaurant is participating in a Meals on Wheels fund-raiser, it certainly makes sense for you to participate—and even offer to ask the winery to participate as well.

"Years ago, I hosted a writer from the *Los Angeles Times* to meet with a winemaker at Wolfgang Puck's Spago restaurant. Chef Puck made an effort to come out of the kitchen to say hello, and it impressed everyone at the table. And he thanked us all for visiting his restaurant. He knows how to play this game," says Paul Wagner.

Of course, you can't do this for every activity that every account mentions to you. That's not anything close to an efficient use of your time. Working with your management team, you should prioritize these activities based on what you think they might accomplish and how they might affect the long-term relationship with the customer. As with everything in the world of sales, your most valuable asset is time, and you will have to prioritize how you spend that time to most effectively reach your sales goals.

Having said that, it would be easy to say that you are very busy making the sales calls themselves and you don't have any time left over to strengthen your relationships with your customers. That would be a big mistake. In the world of consultative selling, you need to make time for both the sales call itself and the building of the closer relationship with your key accounts. You will do both by exceeding expectations in the quantity, quality, and frequency of your interactions with those accounts.

DISCUSSION QUESTIONS

1. How can you become a resource for your key customers?
2. What kinds of research will a small retailer find interesting?
3. How can you share your experience with other customers without violating their confidence?

Chapter Sixteen

Professional Education Development and Your Sales Career Ladder

Our goal in this book has been to provide you the skills and understanding you need to be successful as a wine sales professional. We define success as the ability to create win-win outcomes for the selling and buying organizations as well as for the salesperson. As consultative salespeople, we succeed by helping our customers succeed.

For those of you who pursue the possibilities in sales, you will find a host of opportunities in the wine industry. Though the text has often focused on sales into distribution and wholesale channels along with direct-to-consumer sales, one's career need not be so limited. All wine enterprises, such as wineries, are not only sellers but also buyers of wine-related goods. Who better to sell to wineries as customers than those who have worked in the wine industry? So consider your options in "the world of wine" beyond just the wine product itself.

For those of you who choose not to pursue a sales career, we hope that the skills you have learned in consultative selling, negotiations, and persuasion will be of value as you move forward in your chosen career. The ability to identify and bring different parties together in mutually rewarding win-win relationships is at the heart of entrepreneurship and business leadership.

The skills you develop to become a successful consultative salesperson will serve you well wherever you work, even if your career moves outside the field of sales. The ability to understand the needs of the other person, and to negotiate agreements based on mutual interest, is a vital element in any good executive.

However, there is a second, equally valuable result to your success as a consultative salesperson. The relationships you develop in the world of wine will continue to offer opportunities in the long term. There is no better example of this than the authors' use of key leaders in the world of wine sales to write this book. Our relationships with these leaders were developed over time and allowed us to call on the very best industry experts to illustrate the concepts and examples in this book.

As you continue to explore your career in wine sales, remember how important these relationships are. In many ways, they are the most valuable asset you acquire in your career. We encourage you to think of them not only as relationships but also as an inventory that will continue to appreciate in value as your contacts continue to move up through the ranks in the world of wine.

How valuable are these customers of yours? If we look at some of the brand transactions that have taken place over the past ten years in the world of wine, we can see that major companies have spent hundreds of millions of dollars to purchase brands that own no vineyards and have no winery. What these brands have is an ongoing relationship with an influential group of customers. And those relationships are, in essence, the value of the business. Moreover, when you face the challenge of selling a new and hard-to-understand varietal or placing a new wine on top restaurant lists, you will quickly understand how the relationships you have developed will help your career forever. Protect those relationships, encourage them, and continue to add to your inventory of people in the industry who are in a position to work with you to achieve the goals of whatever company you represent. In the world of wine, consultative relationships are the king of the realm. In the following list, you can see your career ladder as an entry-level wine chain sales merchandiser grow to vice president of sales with increased responsibility and competencies.

Vice President of Sales
  • Industry Competencies
Director of Sales
  • Industry Competencies
General Manager Sales
  • Industry Competencies
Regional Manager
  • Industry Competencies
Territory Manager
  • Industry Competencies
Chain Sales Manager
  • Workplace Competencies
Off-Premise Sales Rep
  • Academic Competencies
Chain Sales Merchandiser
  • Personal Effectiveness Competencies

Opportunities within wine sales are vast, but the skills needed to advance from a sales representative or entry-level position to management usually take wine industry experience. Your first position on this career ladder will bear little resemblance to your next. If becoming a chain sales manager is your goal, a position as an entry-level off-premise sales representative or a chain sales merchandiser can be a strategic first step to get your foot in the door. As an off-premise chain sales representative, you are responsible for building the trust and confidence of the chain store retailer through knowledge of wine products, conducting persuasive presentations, and consultative sales. In chain sales merchandising, you will be working in the fulfillment process of fifteen to twenty-five chain grocery store accounts that are the responsibility of the off-premise chain sales representative. Introducing new products, promoting improved distribution, and ensuring in-store promotion results support the chain sales representative in meeting sales goals.

Eventually, as the chain sales manager, you will find yourself working with the retail wine chain buyer directly to support your off-premise chain representative sales team and chain sales merchandise support staff to sell your wine. Importantly, in this position you will have gained experience in the entry-level work you are instructing others to perform at a high level, offering credibility to your sales management team through experience. The most significant differences are the number of touches you have with the customer and the amount of time spent on each call. While the chain sales manager will likely have from three to five separate meetings with a chain buyer before purchase, the off-premise sales representative may have only had five to ten minutes occasionally with the buyer. Because of this limited time, it's essential to make the most of any opportunity that arises to educate the buyer with your wine and product knowledge.

Frequently we find very talented sales representatives forgo opportunities to advance to the director of sales position or higher. They often confess that they enjoy the freedom and independence in their roles and do not wish to take on the additional responsibilities inherent in the advancements. However, those options can include increased salary and job responsibilities as well as the investment from firms to increase the skills that one needs to master the discipline of sales management.

WINE SCHOOLS AND COLLEGES IN THE UNITED STATES

It is possible to earn a certificate, associate, bachelor's, master's, or doctoral degree in wine-related topics in wine business or management. There are both two-year colleges and four-year universities that offer relevant programs.

Associate Degree Programs

There are several wine-related associate degree options. Students may find a two-year associate of applied science program in the wine business. In these programs, students typically take basic courses in plant science and the wine industry alongside general education courses. They may also complete an internship or cooperative work experience.

Bachelor's Degree Programs

Bachelor's degree programs are available in the wine business as well. Business-focused programs provide an introduction to economic, marketing, and sales principles in the wine industry. Students may be required to complete an internship and a final capstone project before graduation.

Master's Degree Programs

Like bachelor's degree programs, master's degree programs focus heavily on the business aspects of wine. Students may pursue a master of business administration (MBA) or master of science (MS) in wine business and management. These programs combine a core of advanced business courses with wine-specific electives in topics such as global wine distribution and brand management.

Doctoral Degree Programs

Like MBA or MS programs, a PhD in business or hospitality management can include graduate-level wine-related coursework and advanced research opportunities. However, doctoral programs have more extensive requirements and culminate in a final written dissertation and oral defense.

Certificate Programs

Certificate programs typically provide a basic introduction to a particular subject within the field, such as viticulture, enology, or wine marketing. They allow individuals to gain formal education in the field without committing to a full degree program. Some are available in online formats to accommodate the needs of working professionals.

Wine programs confer associate, bachelor's, master's, and doctoral degrees as well as certificates in wine-related topics. The location of the school and the student's career goals are vital considerations when choosing a program. The universities listed in table 16.1 are leaders in the field of wine-related

Table 16.1. Wine Education and Degree Paths at US Colleges and Universities

College/University	Location	Institution Type	Degrees Offered
Napa Valley College	Napa, CA	Two-year, public	Associate
Washington State University	Pullman, WA; Tri-Cities, WA	Four-year, public	Certificate, bachelor's, master's, doctorate
University of California, Davis	Davis, CA	Four-year, public	Certificate, bachelor's, master's
Cornell University	Ithaca, NY	Four-year, private	Bachelor's, master's, doctorate
Sonoma State University	Rohnert Park, CA	Four-year, public	Certificate, bachelor's, master's

business and research offerings. Several wine programs exist domestically in the United States and there are others in Europe and beyond. Table 16.1 shares relevant wine education degree paths at some leading colleges and universities in the United States.

When looking at potential wine schools, it is essential to keep these points in mind:

- Some programs focus more heavily on the scientific aspects of winemaking, while others emphasize the business side of the wine industry. Students should choose between programs based on their academic interests.
- The region and location of the school will dictate the type of grape and wine that students will primarily work with, so prospective students should look for schools in settings where they might like to pursue careers in the future.
- Students who are considering research-based graduate programs should make sure that there are faculty members at the school with whom they share research interests.

CONTINUOUS WINE EDUCATION

Several wine education programs exist to help wine sales professionals continue to develop their opinion leadership in the world of wine. We have highlighted several professional programs below to support your continued wine education.

The Society of Wine Educators

The Society of Wine Educators (SWE) offers several wine certification programs: the Certified Specialist of Wine (CSW) and Certified Wine Educator

(CWE) are the most widely held in the world of wine. Perhaps because of the flexibility that the SWE offers in its self-study educational program, it is internationally recognized and highly regarded in the wine and spirits industry. Through its certification examinations, the Society of Wine Educators validates wine and spirits knowledge. The CSW and CWE are certifications and not certificate programs. Accordingly, successful candidates are entitled to add the appropriate postnominal letters to their professional signature.

The Court of Master Sommeliers

The Court of Master Sommeliers was established to encourage improved standards of beverage knowledge and service in hotels and restaurants. Also known as a wine waiter, a sommelier possesses in-depth knowledge and has a great understanding of wine, which can be put to good use in a variety of settings. Initially, sommeliers worked in hotels and restaurant wine cellars, where they would consult with chefs to decide the best pairing of wines and dishes. Most commonly described as a restaurant professional working with wines, the term *sommelier* is also used by extension for sake, water, or tea. The Court of Master Sommeliers charter is education. The first successful master sommelier examination was held in the United Kingdom in 1969. By April 1977, the Court of Master Sommeliers was established as the premier international examining body of wine. To achieve the title of master sommelier, one must successfully pass all four examinations: (1) Introductory Sommelier Course and Exam, (2) Certified Sommelier Exam, (3) Advanced Sommelier Course and Exam, and (4) Master Sommelier Diploma Exam.

The Wine and Spirit Education Trust

The Wine and Spirit Education Trust, often referred to as WSET, is a British organization that arranges courses and exams in the field of wine and spirits. Headquartered in London, WSET was founded in 1969 and is regarded as one of the world's leading providers of wine education. It grew out of the Wine and Spirit Association's Education Committee and, not unlike the Master Court of Sommeliers, WSET offers four progressive levels of wine education: WSET Level 1 Award in Wines, WSET Level 2 Award in Wines and Spirits, WSET Level 3 Award in Wines, and finally, after successfully testing for the WSET Level 4, a diploma.

The Institute of Masters of Wine

The Institute of Masters of Wine is a professional body with an unsurpassed international reputation. The members, Masters of Wine (MWs), hold the most respected title in the world of wine. MWs have proved their understanding of all aspects of wine by passing the Master of Wine (MW) examination, recognized worldwide for its rigor and high standards. In addition to passing the examination, MWs are required to sign the code of conduct before they are entitled to use the initials MW. The code of conduct requires MWs to act with honesty and integrity and to use every opportunity to share their understanding of wine with others. There are more than 375 Masters of Wine today, working in twenty-nine countries. The membership encompasses winemakers, buyers, journalists, shippers, business owners, consultants, academics, and wine educators. Application for the Master of Wine is selective, with an extensive body of wine knowledge needing to be demonstrated to the institute before an applicant is accepted into the MW program.

IN CONCLUSION

Often we are asked to comment on what the future holds for the sales professional. Change is inevitable, but it is difficult to foresee what advances will take hold and where current practices will remain unchanged. Sales will always be about segmentation, needs generation, value creation, and the delivery on promises we make to our customers. No doubt technology and the Internet will continue to advance and play an ever-growing role in how firms reach and sell to their customers, even at the business-to-business level.

The future of wine sales, we and many others believe, can be thought of as points along a continuum. On one end of the continuum is selling what the consumer considers to be commodities, where convenience and the lowest price always win the business. This form of selling wine will begin to be shaped by the Internet and inside sales teams. Positions within inside sales, e-commerce, and customer relationship management through data analytics influence many wine sales already. On the other end of the continuum is the buyer who is looking to solve complex problems and values the opportunity to work with knowledgeable salespeople in finding ideal solutions. This part of the continuum will remain the domain of the consultative salesperson.

Between these extremes lies a middle ground where technology and lower-cost sales channels can accomplish the same task that has been typically relegated to the sales professional. Southern Glazer's Wine and Spirit Commercial Effectiveness unit is such an attempt, where sales managers work collaboratively with the firm's multiple sales channels to provide customers what they want in the way they prefer acquiring it. Nevertheless, companies like Southern Glazer's must balance efficiency with effectiveness, and some tasks will always require the specialized skills of the sales professional. As Anderson and Narus (1998, p. 59) noted two decades ago, technology, no matter how advanced, "will never replace the salesperson's ability to establish trust with customers, respond to their subtle cues, anticipate their needs, provide personalized service, and create profitable new business strategies with customers." It is the professional consultative salesperson who holds the most promise of migrating consumers from short-term transactional exchanges to long-term relational exchanges.

DISCUSSION QUESTIONS

1. Where can you find wine education near your home?
2. How might a degree of some kind help you advance in your career?
3. If you think about your ultimate career goal, how much will an advanced degree affect your potential success?

CASE STUDY EXERCISE

Todd's Eyes Light Up at the Prospect of a Potential $120,000 Compensation Package

Blue Hills is a custom crush commercial winery based in Richland, Washington. Blue Hills markets specialty wine products designed for corporate gift giving and sells its products primarily in the United States and distributes primarily through third parties. The compensation package for Blue Hills's salespeople consists of a small base salary of $15,000 a year plus a commission on all revenue generated. The average Blue Hills sales manager makes $50,000 to $55,000. During the first year, however, commissions tend to run much lower since much of the salesperson's time is spent learning the business and shadowing other sales managers. The average "rookie" compensation is $32,000 to $35,000.

Todd Bright recently graduated from Washington State University and is working as a bartender. While Todd enjoys his job, he has no intention of making a career out of bartending and is actively looking for career options. Todd recently saw an advertisement announcing an opening for a sales manager for Blue Hills. The advertisement listed multiple benefits and a salary of "up to $120,000 or more." Todd decided to check out the opportunity. He e-mailed his résumé and cover letter to the address provided. Within two days, Sheila Carter, the director of human resources at Blue Hills, called him in for an interview.

The interview went well. Sheila was quite impressed with Todd's education, appearance, and interpersonal skills. She was also impressed that he had taken a course in personal selling as a part of his university coursework. She told him about what the day-to-day job entailed and asked him if he had any questions. Todd asked Sheila if she could tell him more about the compensation package, citing the advertisement that brought Blue Hills to his attention.

Sheila told him that any sales manager "worth his or her salt" should make at least $70,000 their first year. Sheila went on to tell him that it is quite common for a veteran Blue Hills sales manager to earn a six-figure income. She specifically mentioned Jerry Goldman, who in his third year made over $180,000. She offered to put Todd in contact with Jerry. What Sheila did not tell Todd was that Jerry had been brought in from another custom crush company specifically to push Blue Hills's new wine. He was responsible for a considerably larger distributor compared to other sales managers. As an experienced sales professional, his base salary was $40,000, and he was one of only two people charged with selling the new wine.

Todd's eyes opened wide when Sheila mentioned Jerry's compensation. He trusted that Sheila would not offer the reference unless it was true, and he declined the offer to speak with Jerry. Todd was ready to sign on the dotted line if he was offered the job. Sheila telephoned Todd the next day and offered him a new distributor in Seattle, Washington. Todd eagerly accepted.

Todd worked hard while learning the business during his first year in Seattle. To his disappointment, he earned only $34,000 the first year. Looking back, he does not know if Sheila deceived him or if he is just a mediocre sales manager. He is still glad, however, that he took the job. He enjoys his work and his colleagues and is pleased to have a career in sales. He is confident he can earn much more in future years. Todd is especially looking forward to meeting Jerry Goldman, the super sales manager, at the upcoming Blue Hills sales meeting.

DISCUSSION QUESTIONS

1. Have you ever been lured into taking a job only to find out later that the promised benefits were overstated? Describe your experience.
2. Did the Blue Hills advertisement intend to deceive potential sales recruits? How about Sheila's statement to Todd about Jerry's third-year compensation? Is it up to the buyer (in this case Todd) to see through the anticipated puffery in the job advertisement?

References

Anderson, J., & Narus, J. (1998). "Business marketing: Understanding what customers value." *Harvard Business Review, 76*(6), 53–61.

Aziz, A., May, K., & Crotts, J. (2002). "Relationship of Machiavellian behavior and sales performance of stockbrokers." *Psychological Abstracts, 90*, 451–460.

Baldi, L., Peri, M., & Vandone, D. (2013). "Investigating the wine market: A country-level threshold cointegration approach." *Quantitative Finance, 13*(4), 493–503.

Barnes, C. (2015). "Wineries and wine-making industry." Retrieved from https://grapecollective.com/author_pages/christopher-barnes.

Carew, R., & Florkowski, W. (2012). "Wine industry developments in the Pacific Northwest: A comparative analysis of British Columbia, Washington state and Oregon." *Journal of Wine Research, 33*(1), 27–45.

Charters, S., & Michaux, V. (2014). "Strategies for wine territories and clusters." *Journal of Wine Research, 24*(1), 1–4.

Cialdini, R. B. (2001). *Influence: Science and practice* (4th ed.). Boston: Allyn & Bacon.

College Tuition Compare. (2018). Cost of college tuition 2016–2017. Retrieved from https://www.collegetuitioncompare.com.

Constellation Brands. (2017). "Constellation Academy." Retrieved from http://www.academyofwine.com/open-access.

Crotts, J., Alderson-Coppage, C., & Andibo, A. (2001). "Trust-commitment model of buyer-seller relationships." *Journal of Hospitality and Tourism Research, 25*(2), 195–208.

Crotts, J., Aziz, A., & Raschid, A. (1998). "Antecedents of supplier's commitment to wholesale buyers in the international travel trade." *Tourism Management, 19*(2), 127–134.

Crotts, J., Aziz, A., & Upchurch, R. (2005). "The relationship between Machiavellianism and sales performance." *Tourism Analysis, 10*(1), 79–84.

Dororeux, D. (2015). "Use of internal and external sources of knowledge and innovation in the Canadian wine industry." *Canadian Journal of Administrative Sciences, 32*(2), 102–112.

Fisher, R., Ury, W. L., & Patton, B. (1991). *Getting to yes: Negotiating agreement without giving in* (2nd ed.). London: Penguin Books.

Halstead, L. (2016, January 28). *Industry of tomorrow*. Presentation at the Unified Wine & Grape Symposium, Sacramento, CA.

Hunt, S., & Morgan, R. (1995, April). "The comparative advantage theory of competition." *Journal of Marketing, 66*, 1–15.

Ingram, T., LaForge, R., & Schwepker, C. (2007, Fall). "Salespersons ethical decision making: The impact of sales leadership and sales management control strategy." *Journal of Personal Selling and Sales Management, 27*, 301–315.

Jiaqiang, L., Brenard, P., & Plaisent, M. (2013). "Research on Himalayan region wine industrial cluster innovation and management." *Journal of Marketing and Management, 4*(1), 45–58.

MarketLine. (2016). Wine in the United States. Retrieved from https://store.marketline.com/report/mlohme7448--wine-in-the-united-states.

Nielsen, AC. (2012, December 18). Press release: Exploring the alcoholic beverage consumers mindset.

Rackham, N. (1989). *Major account sales strategy*. New York: McGraw Hill.

Richards, K., & Jones, E. (2009, Fall). "Key account management: Adding elements of account fit to an integrative theoretical framework." *Journal of Personal Selling and Sales Management, 29*, 305–320.

Roberto, M. (2005). *Robert Mondavi and the wine industry*. Cambridge, MA: Harvard Business School (Case No. 9-302-102).

Shell, R., and M. Moussa (2007). *The art of woo: Using strategic persuasion to sell your ideas*. New York: Penguin.

Teague, L. (2015). *Wine in words*. New York: Rizzoli.

Thach, L., & Chang, K. (2015, November 11). 2015 survey of American wine consumer preferences. Retrieved from http://www.winebusiness.com/news/?go=getArticle&dataid=160722.

Thomas, L., Gómez, M. I., Gerling, C. J., & Mansfield, A. K. (2014). "The effect of tasting sheet sensory descriptors on tasting room sales." *International Journal of Wine Business Research, 26*(1), 61–72.

Turner, J., & Turner, T. (1999). "Determinants of intra-firm trust in buyer-seller relationships in the international travel trade." *International Journal of Contemporary Hospitality Management, 11*(2/3), 116–123.

US Treasury Department, Alcohol and Tobacco Taxes and Trade Bureau. (2015). "Wine statistics." Retrieved from https://ttb.gov/wine/wine-stats.shtml.

———. (2016). "Wine statistics." Retrieved from https://ttb.gov/wine/wine-stats.shtml.

Voronov, M., DeClercq, D., & Hinings, C. (2013). "Conformity and distinctiveness in a global institutional framework: The legitimation of Ontario fine wines." *Journal of Management Studies, 50*(4), 607–645.

Wagner, P., J. Olsen, and L. Thach. (2011). *Wine marketing and sales.* (2nd ed.). San Francisco: Board and Bench.

Washington Association of Wine Growers. (2015). "2015 grape production report." Retrieved from https://greatnorthwestwine.com.

Wine and Spirits Wholesale of America Association. (2017). "By the numbers: The wine and spirit industry." Retrieved from http://www.wswa.org/facts-data.

Wine Institute. (2016, February 25). "California wine exports set record in 2015: Worldwide demand grows, despite strong dollar." *Wine Institute Website.* Retrieved from http://www.wineinstitute.org/resources/pressroom/02252016.

Index

Note: Figures and tables are indicated by "f" and "t," respectively, following the page number.

account management, 39–41, 46, 79
account ownership, 34
agenda, for sales calls, 81–82, 90–95
AIDA (awareness, interest, desire, action) sales model, 4
Alcohol and Tobacco Tax and Trade Bureau, 15
alliance sales, 36
Anderson, J., 181
appearance and demeanor, 83, 87–88, 90
associate degree programs, 177–78
Australia, 4

bachelor's degree programs, 178
BATNA (best alternative to a negotiated agreement), 26–27, 37
Bell, Joshua, 89–90
best customers, 48–49
brand: building, through wines sales, 28–29, 42; emotional/personal connections to, 72; four Ps of, 161; management of, 27–28, 31, 41; placement as factor in, 41, 161–64; pricing as factor in, 43; promotion

as factor in, 164–67; relationship as component of, 176
Breakthru Beverage, 5, 36
buyers, types of, 14–15. *See also* consumers; organizational buyers; panel of buyers
buyer's remorse, 86
buying signals, 125–26, 131–32

call opening, 74, 87–96; agenda, 90–95; first impressions, 87–88, 90; tips on influence, 88–90
case cards, 166
certificate programs, 178
Certified Specialist of Wine, 179–80
Certified Wine Educator, 179–80
chain buyers, 31–33, 42–43
chain sales managers, 176–77
Chang, K., 51–52
channel mix, 29
channels. *See* sales channels
Chateau Ste. Michelle, 36
checking for acceptance, 92–93, 118, 130, 143, 148
China, 4, 26

Cialdini, Robert, 88, 90
closing the sale: contracts and vendor agreements, 135–36; in DTC sales, 55–56, 75; how to close, 126–28; next steps in conjunction with, 129–30; in organizational sales, 125–36; preparing for, 131; responding to rejection, 133–34 (*see also* lack of interest, selling to); response to concerns in, 132–33; tips for, 133; when to close, 125–26
Coca-Cola, 60
cold boxes, 164–65
commissions, sales, 9, 62–63
commitment, in sales relationships, 89
compensation, 8–9, 61–63, 182–83
compromise, 139
consistency, in sales relationships, 89
Constellation Brands, 25
consultative sales process: agenda for, 81–82, 90–95; benefits of, for winery, 14; considerations in approaching, 67–69; elements and flow of, 73–75; fundamentals of, 69–73; long-term relationships as goal of, 13, 21, 35, 79, 169–73, 175–76; for organizational buyers, 65; overview of, 35–36; value added through, 24–25, 35, 81, 107t, 110–13, 153, 171–73
consumers: motivations for buying, 51–52, 52f, 57; needs of, 53–54; negative feelings/experiences of, 57–61; taste preferences of, 52f, 53; types of, 25t; value for, 24
contacts, management of, 45–46
contracts, 135–36
control states, 135
Cornell University, 57
coupons, 166
Court of Master Sommeliers, 180
credibility, 4, 73, 83, 88
CRM (customer relationship management) programs, 44, 48

customer needs: acknowledging, 114–16; assessing customer's circumstances, 100–102; assessing customer's strategy, 99–100; behind false impressions, 145–46; behind the expressed need, 102–3, 117; in consultative sales process, 69–71; in DTC sales, 53–54; gap analysis of, 105–7; importance of determining, 4; of large-volume clients, 84; open vs. closed probes, 103–5; in organizational sales, 97–108; probing for, 53–54, 74–75, 97–99, 103–5, 154–55, 156t, 170

data mining, 171
derived demand, 79, 82–83
direct-to-consumer (DTC) sales, 51–63; compensation in, 61–63; costs of, 30; sales process, 53–56; small wineries' use of, 28, 41; supplementing, 41–42; things to avoid, 57–61; volume of sales from, 41, 51
discounts, 19–20, 43
distributors and distribution: embedded sales reps for, 10; laws and regulations regarding, 5; retailers' relationships to, 135–36; worldwide, 4–6
doctoral degree programs, 178
DTC. *See* direct-to-consumer (DTC) sales

education. *See* professional education development
E. J. Gallo. *See* Gallo Winery
end-aisle displays, 80, 101, 166–67
ethics, 23, 73, 90
Europe, 5, 6
exclusivity, 71
expertise, 4, 36, 60, 115, 123

fact sheets, 94, 112, 144, 165–66
fair trade certification, 157
Falgueras, Meritxell, *59*, 60n

false impressions, buyers', 139–41, 144–46

features and benefits, 57, 107t, 110–13, 117, 155, 155t, 157

first impressions, 87–88, 90

Fitzgerald, F. Scott, 32

floor stacks, 80, 101, 162, 166–67

follow-ups, 44, 56

food-and-wine pairings, 72, 80–81, 95, 101, 103, 118, 155, 162, 163, 166, 167

France, 4

franchise states, 10, 135–36

free shipping, 17

Gallo Winery, 36, 164

gaming industry, 48

gap analysis, 72, 80, 93, 99, 105–7, 105f, 106f

Germany, 4, 26

greeting, 53

guests, member benefits for, 48–49

HACCP certification, 85, 157

Hemingway, Ernest, 32

holiday promotions, 42

hospitality companies, 48–49

inbound call centers, 9

Institute of Masters of Wine, 180–81

Internet, 28, 30, 33, 35

inventory management, 7

ISO certification, 85, 157

Italy, 4

Kendall Jackson, 36

key account managers, 10

key accounts. *See* large-volume customers

Kramer, Matt, 6

lack of interest, selling to, 75, 153–58

large-volume customers: impact of, 5–6; sales process of, 32, 84–86; selling to, 31–33, 83

laws and regulations: common issues involving, 16–17; contracts and vendor agreements, 135–36; country-specific, 5; federal and state, 15–16; governing sales, 3

letters of agreement, 136

likeability, of salespeople, 88

listening, 7, 53, 55, 70, 94, 100, 102, 114, 170

manipulative tactics, 137

markups, 18

master's degree programs, 178

Masters of Wine, 180–81

McCarthy, Sharon, 120n

Mediterranean countries, 26

member benefits for guests, 48–49

merchandising, 161–68; defined, 162; placement as component of, 161–64; promotion as component of, 164–67

mystery shoppers, 90, 110

Narus, J., 181

neck-hangers, 162, 166

negative responses, overcoming of, 55

negotiation: with organizational buyers, 85, 137–52; probing questions in, 141–42; response to concerns in, 55, 75, 137–52; role of, in sales situations, 55; timing of, in sales calls, 93–94; types of objections in, 139–40, 142t; win-win outcomes as goal of, 138–39. *See also* BATNA

nonfranchise states, 135–36

off-premise sales and accounts, 15, 29, 176–77

on-premise sales and accounts, 15, 28–29, 41–42, 80–81

open-ended questions, 53–54, 103–5

organizational buyers: chain buyers, 31–33, 42–43; considerations in approaching, 67–69; derived demand, 82–83; distinctive features of, 82–86; goal of, 79; needs of,

69–71; relationship building with, 169–73; risk reduction strategies of, 67; size and frequency of orders, 83; sophistication of, 83–84

outweighing a buyers' concerns, 148–49

pairings. *See* food-and-wine pairings
panel of buyers, 83
partnering sales, 36
Peynaud, Emile, 60
placement, 41–42, 161–64
point-of-sale (POS) merchandising, 165–67
precall research, 74, 77–86; anticipation of buyer concerns, 152, 155, 157; buyer information, 79; competitor information, 80; importance of, 77–79; inventory and market information, 80; for organizational buyers, 82–86; sales call plan based on, 81–82; store/restaurant information, 80, 81; tips for, 78–79
pricing: buyers' considerations concerning, 70; conflicts in, 30, 43; considerations in, 17–21; discounts in, 19–20, 43; perceived value in relation to, 19; regulations on, 16; in restaurants, 18, 20–21, 30; three-tier system of, 17–18, 18t; transactional sales and, 34–35
production of wine, worldwide, 4–6
professional education development, 177–81, 179t
Prohibition, 16
promotion of wines, 164–67
proof devices, 112, 118, 143–44
Puck, Wolfgang, 173
purchase incentives, 70–71
purchasing partnerships, 10

rapport building, 54. *See also* relationship building
ratings, 57, 70
reciprocity, in sales relationships, 88–89
refocusing, of buyers' concerns, 147–48
regulations. *See* laws and regulations

relationship building, 13, 75, 169–73, 175–76. *See also* rapport building
requests for proposals (RFPs), 82–83, 85, 94
restaurants: placement of wines in, 41–42, 163–64; pricing in, 18, 20–21, 30; retail-store competition with, 30, 43, 71; selling strategies in, 20, 28–29
retailers: distributors' relationships to, 135–36; restaurants' competition with, 30, 43, 71; sales process of, 32; tasting rooms' competition with, 30, 31
Robert Parker's Wine Advocate, 57

salary, 9
sales call reports, 44
sales call technology, 45–46
sales channels: conflicting, 29–31, 43, 71; integration of, 33–34; intensive vs. selective, 28f, 29; management strategies for, 27–34
sales force, 39–50; account management, 39–41, 46; analysis of results, 47; "best customer" determinations, 48–49; extending member benefits, 48–49; sales call reports, 44; technology for, 45–46; territory management, 41–43
sales managers, 39–40
sales orders, 45
salespeople: appearance and demeanor of, 83, 87–88, 90; career paths of, 9–10, 175–77; career rewards for, 8–9; characteristics of successful, 7–8, 10–11, 175; embedded in distributors, 10; future prospects of, 181–82; importance of, 13, 14, 23, 35–36, 36–37; influence of, 88–90; probing skills of, 97–99. *See also* sellers and selling
sales pitch, 54–55
scarcity, 89
sellers and selling: AIDA model for, 4; BATNA as factor in, 26–27,

37; factors in, 3; goal of, 13; to a lack of interest, 75, 153–58; to mismatched customers, 134; process of, 14, 53–56; responding to buyer concerns, 55, 75, 132–33, 137–52; restaurant strategies for, 20, 28–29; shortcomings of, 140–41, 147–51; SMART objectives for, 44; tailoring of, 15; techniques for, 72–73; things to avoid, 57–61, 119–20; time management and prioritization, 79, 104, 134, 157, 173; types of, 34–36; value for, 23. *See also* consultative sales process; salespeople

shelf space, 163–65

shelf-talkers, 94, 162, 165, 166

Shell, Richard, 78–79

skepticism, buyers', 139–44

SMART (specific, measurable, achievable, relevant, time-bound) selling objectives, 44

smartphones, 20, 107, 165

social proof, 89

Society of Wine Educators, 179–80

sommeliers, 10, 19, 180

Southern Glazer's Wines and Spirits, 5, 36, 181

spreadsheets, 45t

Stanton, Andrew, 122

state laws and regulations, 5, 16–17, 135–36

status quo, challenges to, 72–73, 103, 153

stories, as selling tool, 57–58, 72, 119–23, 172

suggested retail price, 17

supermarkets. *See* chain buyers

support for customers, 71, 74, 109–24; how to provide, 114–18; preparation for providing, 119; providing information, 109–13; when to provide, 114

sweepstakes, 17

table tents, 162, 165, 166

taste, 51–53, 52f

tasting notes, 57–58, 61

tasting rooms. *See* winery tasting rooms

Teague, Lettie, 53

technology, 45–47, 181–82

territory management, 41–43

Thach, L., 51–52

thought leadership, 72

three-tier system, 16, 17–18, 120

transactional sales, 34–35

trust, 4, 7–8, 73, 88, 100

United States: laws and regulations in, 5, 15–16, 135–36; production and consumption of wine in, 4–5; three-tier system in, 16

US Department of the Treasury, 15

value: added through consultative sales process, 24–25, 35, 81, 107t, 110–13, 153, 171–73; for buyers, 24, 92, 128; defined, 24; perceptions of, 19, 24; for sellers, 23

vendor agreements, 135–36

waiter cards, 166

wine-and-food pairings. *See* food-and-wine pairings

Wine and Spirit Education Trust, 180

wine clubs, 40, 46, 47, 48, 49

wine industry: career paths in, 9–10, 175–77; future of, 181–82; professional education development for, 177–81, 179t

wineries: benefits of consultative sales process for, 14; DTC used by, 28, 41; mistakes of, 60–61

winery tasting rooms: compensation in, 9, 61–63; importance of salespeople in, 23, 25; retail-store competition with, 30, 31; sales process in, 53–56; selling strategies in, 31; as significant sales channel, 41

Wine Spectator, 57

win-win outcomes, 23, 72, 80, 88, 138–39, 175

Young's Market Co., 5, 36

About the Authors

Paul Wagner is the founder of Balzac Communications & Marketing and has taught wine marketing and sales at Napa Valley College for more than twenty-five years. He has been a regular columnist for *Vineyard & Winery Management* magazine and is an international wine judge, a founding member of the Academy of Wine Communications, and an inductee of the Spadarini della Castellania di Soave. In 2009, Mr. Wagner was honored with a Life Dedicated to Wine Award at the Feria Nacional del Vino (FENAVIN) in Spain, and in 2018 he was given the Award of Merit by the American  Wine Society. In 2018, Napa Valley College honored him with the McPherson Distinguished Teaching Award. He was the lead author of *Wine Marketing & Sales: Success Strategies for a Saturated Market* (Wine Appreciation Guild). The book, now in its third edition, won the Gourmand International Award for best international wine book for professionals.

 **John C. Crotts**, PhD, is a professor of hospitality and tourism management in the School of Business at the College of Charleston in Charleston, South Carolina. His research and teaching encompass the areas of economic psychology, sales and negotiation strategies, and management of cooperative alliances. He is the coauthor of *Selling Hospitality: A Situational*

Approach (Delmar/Thompson Publishing). He also serves as the North American regional editor of *Tourism Management* and is on the editorial boards of over fourteen tourism research journals, including the *Journal of Travel Research*, the *Journal of Business Research*, and the *Journal of Travel and Tourism Marketing*. He is also affiliated with the Services Research Institute of Virginia Tech University and the Tourism Center at MCI-Innsbruck, Austria.

Byron Marlowe, PhD, clinical assistant professor, coordinates Washington State University's Wine and Beverage Business Management Program and is a faculty member in the Carson College of Business, School of Hospitality Business Management at WSU Tri-Cities. Byron is an International Business Fellow in the Carson College of Business and holds several international visiting lecturer/professor positions at the University of Applied Sciences, Hochschule Harz, Germany; Institut Paul Bocuse, France and China; and Castello Sonnino Field Study, Italy, for his expertise in wine tourism as well as food and beverage management. Byron's teaching background is from Southern Oregon University, where he was previously senior instructor and hospitality and tourism program coordinator in the School of Business, and Le Cordon Bleu College, where he was the lead instructor in the culinary management program.